FREE

Free Study Tips DVD

In addition to the tips and content in this guide, we have created a FREE DVD with helpful study tips to further assist your exam preparation. **This FREE Study Tips DVD provides you with top-notch tips to conquer your exam and reach your goals.**

Our simple request in exchange for the strategy-packed DVD is that you email us your feedback about our study guide. We would love to hear what you thought about the guide, and we welcome any and all feedback—positive, negative, or neutral. It is our #1 goal to provide you with top quality products and customer service.

To receive your **FREE Study Tips DVD**, email freedvd@apexprep.com. Please put "FREE DVD" in the subject line and put the following in the email:

a. The name of the study guide you purchased.

b. Your rating of the study guide on a scale of 1-5, with 5 being the highest score.

c. Any thoughts or feedback about your study guide.

d. Your first and last name and your mailing address, so we know where to send your free DVD!

Thank you!

The AFOQT Tutor

AFOQT Study Guide 2020-2021 Prep and Practice Test Questions for the Air Force Officer Qualifying Test
[Includes Detailed Answer Explanations]

APEX Test Prep

Table of Contents

Test Taking Strategies

1. Reading the Whole Question

A popular assumption in Western culture is the idea that we don't have enough time for anything. We speed while driving to work, we want to read an assignment for class as quickly as possible, or we want the line in the supermarket to dwindle faster. However, speeding through such events robs us from being able to thoroughly appreciate and understand what's happening around us. While taking a timed test, the feeling one might have while reading a question is to find the correct answer as quickly as possible. Although pace is important, don't let it deter you from reading the whole question. Test writers know how to subtly change a test question toward the end in various ways, such as adding a negative or changing focus. If the question has a passage, carefully read the whole passage as well before moving on to the questions. This will help you process the information in the passage rather than worrying about the questions you've just read and where to find them. A thorough understanding of the passage or question is an important way for test takers to be able to succeed on an exam.

2. Examining Every Answer Choice

Let's say we're at the market buying apples. The first apple we see on top of the heap may *look* like the best apple, but if we turn it over we can see bruising on the skin. We must examine several apples before deciding which apple is the best. Finding the correct answer choice is like finding the best apple. Although it's tempting to choose an answer that seems correct at first without reading the others, it's important to read each answer choice thoroughly before making a final decision on the answer. The aim of a test writer might be to get as close as possible to the correct answer, so watch out for subtle words that may indicate an answer is incorrect. Once the correct answer choice is selected, read the question again and the answer in response to make sure all your bases are covered.

3. Eliminating Wrong Answer Choices

Sometimes we become paralyzed when we are confronted with too many choices. Which frozen yogurt flavor is the tastiest? Which pair of shoes look the best with this outfit? What type of car will fill my needs as a consumer? If you are unsure of which answer would be the best to choose, it may help to use process of elimination. We use "filtering" all the time on sites such as eBay® or Craigslist® to eliminate the ads that are not right for us. We can do the same thing on an exam. Process of elimination is crossing out the answer choices we know for sure are wrong and leaving the ones that might be correct. It may help to cover up the incorrect answer choice. Covering incorrect choices is a psychological act that alleviates stress due to the brain being exposed to a smaller amount of information. Choosing between two answer choices is much easier than choosing between all of them, and you have a better chance of selecting the correct answer if you have less to focus on.

4. Sticking to the World of the Question

When we are attempting to answer questions, our minds will often wander away from the question and what it is asking. We begin to see answer choices that are true in the real world instead of true in the world of the question. It may be helpful to think of each test question as its own little world. This world may be different from ours. This world may know as a truth that the chicken came before the egg or may assert that two plus two equals five. Remember that, no matter what hypothetical nonsense may be in the question, assume it to be true. If the question states that the chicken came before the egg, then choose your answer based on that truth. Sticking to the world of the question means placing all of our biases and assumptions aside and relying on the question to guide us to the correct answer. If we are simply looking for answers that are correct based on our own judgment, then we may choose incorrectly. Remember an answer that is true does not necessarily answer the question.

5. Key Words

If you come across a complex test question that you have to read over and over again, try pulling out some key words from the question in order to understand what exactly it is asking. Key words may be words that surround the question, such as *main idea, analogous, parallel, resembles, structured,* or *defines.* The question may be asking for the main idea, or it may be asking you to define something. Deconstructing the sentence may also be helpful in making the question simpler before trying to answer it. This means taking the sentence apart and obtaining meaning in pieces, or separating the question from the foundation of the question. For example, let's look at this question:

> Given the author's description of the content of paleontology in the first paragraph, which of the following is most parallel to what it taught?

The question asks which one of the answers most *parallels* the following information: The *description* of paleontology in the first paragraph. The first step would be to see *how* paleontology is described in the first paragraph. Then, we would find an answer choice that parallels that description. The question seems complex at first, but after we deconstruct it, the answer becomes much more attainable.

6. Subtle Negatives

Negative words in question stems will be words such as *not, but, neither,* or *except.* Test writers often use these words in order to trick unsuspecting test takers into selecting the wrong answer—or, at least, to test their reading comprehension of the question. Many exams will feature the negative words in all caps (*which of the following is NOT an example*), but some questions will add the negative word seamlessly into the sentence. The following is an example of a subtle negative used in a question stem:

> According to the passage, which of the following is *not* considered to be an example of paleontology?

If we rush through the exam, we might skip that tiny word, *not,* inside the question, and choose an answer that is opposite of the correct choice. Again, it's important to read the question fully, and double check for any words that may negate the statement in any way.

7. Spotting the Hedges

The word "hedging" refers to language that remains vague or avoids absolute terminology. Absolute terminology consists of words like *always, never, all, every, just, only, none,* and *must.* Hedging refers to words like *seem, tend, might, most, some, sometimes, perhaps, possibly, probability,* and *often.* In some cases, we want to choose answer choices that use hedging and avoid answer choices that use absolute terminology. It's important to pay attention to what subject you are on and adjust your response accordingly.

8. Restating to Understand

Every now and then we come across questions that we don't understand. The language may be too complex, or the question is structured in a way that is meant to confuse the test taker. When you come across a question like this, it may be worth your time to rewrite or restate the question in your own words in order to understand it better. For example, let's look at the following complicated question:

> Which of the following words, if substituted for the word *parochial* in the first paragraph, would LEAST change the meaning of the sentence?

Let's restate the question in order to understand it better. We know that they want the word *parochial* replaced. We also know that this new word would "least" or "not" change the meaning of the sentence. Now let's try the sentence again:

> Which word could we replace with *parochial,* and it would not change the meaning?

Restating it this way, we see that the question is asking for a synonym. Now, let's restate the question so we can answer it better:

> Which word is a synonym for the word *parochial*?

Before we even look at the answer choices, we have a simpler, restated version of a complicated question.

9. Predicting the Answer

After you read the question, try predicting the answer *before* reading the answer choices. By formulating an answer in your mind, you will be less likely to be distracted by any wrong answer choices. Using predictions will also help you feel more confident in the answer choice you select. Once you've chosen your answer, go back and reread the question and answer choices to make sure you have the best fit. If you have no idea what the answer may be for a particular question, forego using this strategy.

10. Avoiding Patterns

One popular myth in grade school relating to standardized testing is that test writers will often put multiple-choice answers in patterns. A runoff example of this kind of thinking is that the most common answer choice is "C," with "B" following close behind. Or, some will advocate certain made-up word patterns that simply do not exist. Test writers do not arrange their correct answer choices in any kind of pattern; their choices are randomized. There may even be times where the correct answer choice will be the same letter for two or three questions in a row, but we have no way of knowing when or if this might happen. Instead of trying to figure out what choice the test writer probably set as being correct, focus on what the *best answer choice* would be out of the answers you are presented with. Use the tips above, general knowledge, and reading comprehension skills in order to best answer the question, rather than looking for patterns that do not exist.

FREE DVD OFFER

Achieving a high score on your exam depends not only on understanding the content, but also on understanding how to apply your knowledge and your command of test taking strategies. **Because your success is our primary goal, we offer a FREE Study Tips DVD, which provides top-notch test taking strategies to help you optimize your testing experience.**

Our simple request in exchange for the strategy-packed DVD is that you email us your feedback about our study guide.

To receive your **FREE Study Tips DVD**, email freedvd@apexprep.com. Please put "FREE DVD" in the subject line and put the following in the email:

    a. The name of the study guide you purchased.

    b. Your rating of the study guide on a scale of 1-5, with 5 being the highest score.

    c. Any thoughts or feedback about your study guide.

    d. Your first and last name and your mailing address, so we know where to send your free DVD!

Introduction to the AFOQT

Function of the Test

The United States Air Force administers the Air Force Officer Qualifying Test (AFOQT), a standardized test that measures the knowledge and aptitude of prospective candidates for Air Force Officer training programs. The exam consists of twelve subtests, which together evaluate a test taker's proficiency in a variety of domains deemed important to success in Air Force career paths. Subtests include verbal, mathematical, and physical science knowledge, as well as reading comprehension, aviation understanding, and other specific skills and aptitudes such as situational judgement and instrument comprehension. The AFOQT is used as a component of the admissions process to Air Force Officer training programs. Within the United States Air Force, the test is part of the Pilot Candidate Selection Method (PCSM) score and is used to determine the qualifications of candidates for Pilot, Combat Systems Officer (CSO), and Air Battle Manager (ABM) training. All Air Force Officer students receiving a scholarship or in the Professional Officer Course (POC) must take the AFOQT.

Current and prospective candidates for the United States Air Force take the AFOQT. In the summer following sophomore year, students in the Air Force ROTC program take the exam prior to their field training.

Test Administration

Air Force ROTC programs in the United States administer the AFOQT on college campuses. Military recruiters administer the exam at Military Entrance Processing facilities. Candidates do not incur a cost to take the AFOQT, but they are required to arrange to take the exam through their ROTC program, commanding officer, or military recruiter, depending on their personal circumstances. Set retesting rules are not in place; instead, the ability to take the test again depends on permissions afforded by each specific program.

Test Format

The AFOQT is filled out by hand on paper and scored via machine. The total time allotted for the exam, including breaks and administration time is 4 hours and 47.5 minutes. Of that, 3 hours and 36.5 minutes is dedicated to answering questions in the actual test sections. The AFOQT consists of twelve subtests, each with multiple-choice questions containing four or five answer choices. The subtests, in order, are as follows: verbal analogies, arithmetic reasoning, word knowledge, math knowledge, reading comprehension, situational judgment, a self-description inventory, physical science, table reading, instrument comprehension, block counting, and aviation information.

A summary of the subtests, including the number of questions and allotted time is as follows:

Subtest	Items	Time (min.)
Verbal Analogies	25	8
Arithmetic Reasoning	25	29
Word Knowledge	25	5
Math Knowledge	25	22
Reading Comprehension	25	39
Situational Judgement	50	35
Self-Description Inventory	240	45
General Science	20	10
Table Reading	40	7
Instrument Comprehension	25	5
Block Counting	30	4.5
Aviation Information	20	8
TOTAL	**545**	**3 hours, 36.5 minutes**

Scoring

Scores are calculated from the number of correct answers. No penalty is incurred for incorrect responses, so it is wise for test takers to guess and select an answer choice even if they are uncertain of the correct answer. The test taker and the Air Force receive score reports that contain the test taker's composite scores, which are calculated from different combinations of the scores from various subtests. The "Pilot" composite score is calculated from the results of the math knowledge, instrument comprehension, table reading, and aviation information subtests. The "Combat Systems Officer" composite score is aggregated from the results of the word knowledge, math knowledge, table reading, and block counting sections. The "Air Battle Manager" composite score is the sum of the verbal analogies, math knowledge, table reading, instrument comprehension, block counting, and aviation information subtests. Other composite scores exist as well. A percentile score from 0 to 99 in each of the composite categories is provided on the test taker's score report.

The AFOQT does not have a set passing score. Instead, the minimum scores required vary between different programs or jobs to which the test taker is applying. For instance, a test taker wishing to become a pilot will likely need a much higher percentile score (so he or she must outperform the majority of test takers), particularly in the Pilot composite score, than one seeking to become an officer.

Verbal Analogies

Verbal Analogies

The verbal analogies section of the AFOQT is designed to test your knowledge of word definitions as well as their relationship to one another and as a pair to another pair of words. Making analogies is part of an important cognitive process that acts as the basis of metaphor and association. Analogical thinking is used in problem solving, creative thinking, argumentation, invention, communication, and memory, among other intellectual operations. Verbal analogies are a way to determine associations between objects, their signifying words, and each word's specific connotations. Learning verbal analogies is valued among many standardized exams because of its usefulness in language and learning, especially among figurative language such as metaphors, similes, and allegories.

Question Format

Below we will look at the different types of analogies, but it's also helpful to know what format the analogy section uses and how best to answer the questions. The question will give you either two words or three words to work with. Let's start with an example with only two words:

Cat is to **mammal** as

a. **shoe** is to **foot**.
b. **kitten** is to **feline**.
c. **dolphin** is to **amphibian**.
d. **lizard** is to **reptile**.
e. **lamp** is to **bedroom**.

This type of question requires you to carefully study the relationship between the first pair of words. The first pair of words have the relationship of **category and type.** We see that in the world of mammals, a *cat* is a **type** of mammal, or can be considered a mammal. The word "cat" falls within the classification of "mammal." Now when we look at the answer choices, we have to determine that the first word falls within an appropriate category of the second word. "Shoe" is not a category of "foot," so Choice *A* is incorrect. "Feline" is a synonym of "cat," or is part of the cat family, so this is not the best answer. Choice *B*'s "kitten" and "feline" are too close to each other in meaning. Choice *C* is incorrect because a dolphin does not fall under the class of "amphibian." Rather, it is a mammal. Choice *D* looks like the correct answer. "Lizard" is a type of "reptile" and falls under the category of "reptile." This pair of words has the same relationship as the original pair. Finally, Choice *E* is incorrect. A lamp may reside in the bedroom, but the association is too generalized for this question.

Now let's look at a question with three words to start off with:

Heat is to **scorching** as **cold** is to

a. burning.
b. freezing.
c. melting.
d. ice.
e. Alaska.

In the questions where we are given three terms, we still need to identify the relationship between the first pair of words. Here, we see that this type of analogy is a **degree of intensity**. We are given the word "hot," and we know that "scorching" means "really, really hot." Let's look at what cold's intensity looks like. Choices *A* and *C*, burning and melting, are the opposite of an intensity of cold, so these are incorrect. Choice *B* looks like a good answer. "Freezing" is an intensity of cold, so let's mark this as correct. Choice *D*, ice, is an object that is really cold, and Choice *E*, Alaska, is a place that is really cold. These could be in the running. However, to narrow down between "freezing," "ice," and "Alaska," it's important to look at the closest analogy to the word "scorching." The word is an adjective, not a noun or proper noun, as ice and Alaska are. Since "scorching" and "freezing" both occupy the same part of speech, let's choose "freezing," Choice *B*, as the correct answer.

Types of Analogies

In its most basic form, an analogy compares two different things. An analogy is a pair of words that parallels the situation or relationship given in another pair of words. The table below give examples of the different types of analogies you may see on the AFOQT:

Type of Analogy	Relationship	Example
Synonym	The pair of words are alike in meaning.	**Happy** is to **joyous** as **sad** is to **somber**.
Antonym	The pair of words are opposite in meaning.	**Lucky** is to **unfortunate** as **victorious** is to **defeated**.
Part to Whole	One word stands for a whole, and the other word stands as a part to that whole.	**Chapter** is to **novel** as **pupil** is to **eye**.
Category/Type	One thing belongs in a category of another thing.	**Screwdriver** is to **tool** as **apartment** is to **dwelling**.
Object to Function	A pair depicting a tool and the use of that tool.	**Shovel** is to **dig** as **oven** is to **bake**.
Degree of Intensity	The pair of words shows a difference in degree.	**Funny** is to **hysterical** as **interest** is to **adoration**.
Cause and Effect	The pair shows that one word is created by the other word.	**Hard work** is to **success** as **privilege** is to **comfort**.
Symbol and Representation	A word and its representation in the context of a culture.	**Rose** is to **love** as **flag** is to **patriotism**.
Performer to Related Action	A person and their related action.	**Professor** is to **teach** as **doctor** is to **heal**.

Practice Questions

1. **Chapter** is to **book** as
   a. Book is to story.
   b. Fable is to myth.
   c. Paragraph is to essay.
   d. Dialogue is to play.
   e. Story is to tale.

2. **Dress** is to **garment** as
   a. Diesel is to fuel.
   b. Month is to year.
   c. Suit is to tie.
   d. Clothing is to wardrobe.
   e. Coat is to winter.

3. **Car** is to **garage** as **plane** is to
   a. Sky
   b. Passenger.
   c. Airport.
   d. Runway.
   e. Hanger.

4. **Trickle** is to **gush** as
   a. Bleed is to cut.
   b. Rain is to snow.
   c. Tepid is to scorching.
   d. Sob is to sniffle.
   e. Ocean is to river.

5. **Acne** is to **dermatologist** as **cataract** is to
   a. Psychologist.
   b. Ophthalmologist.
   c. Otolaryngologist
   d. Otologist.
   e. Orthopedist.

6. **Jump** is to **surprise** as
   a. Run is to walk.
   b. Spook is to scare.
   c. Chuckle is to joke.
   d. Hop is to bunny.
   e. Sadness is to cry.

7. **Knife** is to **slice** as **fork** is to
    a. Cut.
    b. Spear
    c. Mouth.
    d. Spoon.
    e. Eat.

8. **Eat** is to **ate** as **spin** is to
    a. Thread.
    b. Spinned.
    c. Spinning.
    d. Spun.
    e. Spins.

9. **Obviate** is to **preclude** as
    a. Exclude is to include.
    b. Conceal is to avert.
    c. Pontificate is to ponder.
    d. Appease is to placate.
    e. Ostentatious is to poignant.

10. **Acre** is to **area** as **fathom** is to
    a. Depth.
    b. Angle degree.
    c. Wind speed.
    d. Ship.
    e. Width.

11. **Green** is to **blue** as **orange** is to
    a. Red.
    b. Yellow.
    c. Purple.
    d. Green.
    e. Blue.

12. **Peel** is to **orange** as
    a. Fur is to bear.
    b. Shed is to snake.
    c. Peel is to sunburn.
    d. Shell is to coconut.
    e. Fuzz is to peach.

13. **Sycophant** is to **flattery** as **raconteur** is to
    a. Heritage.
    b. Philosophies.
    c. Idioms.
    d. Artwork.
    e. Anecdotes.

14. **Viscous** is to **runny** as
   a. Somber is to merciful.
   b. Vacuous is to nostalgic.
   c. Plunder is to laud.
   d. River is to stream.
   e. Obscure is to unequivocal.

15. **Nadir** is to **zenith** as **valley** is to
   a. Depression.
   b. Pinnacle.
   c. Climb.
   d. Rise
   e. Slope.

16. **Pundit** is to **expertise** as **scholar** is to
   a. Study.
   b. Learning.
   c. Novice.
   d. Teacher.
   e. Erudition.

17. **Blueprint** is to **architect** as
   a. Stethoscope is to doctor.
   b. Model is to train.
   c. Lathe is to craftsman.
   d. Outline is to drawing.
   e. Score is to composer.

18. **Evidence** is to **detective** as **gold** is to
   a. Jeweler.
   b. Pan.
   c. Prospector.
   d. Magnet.
   e. Archeologist.

19. **Odometer** is to **distance** as **caliper** is to
   a. Pressure.
   b. Thickness.
   c. Wind.
   d. Brake.
   e. Body.

20. **Radish** is to **vegetable** as
   a. Garbanzo is to legume.
   b. Pineapple is to berry.
   c. Lettuce is to spinach.
   d. Cucumber is to salad.
   e. Citrus is to fruit.

21. **Adore** is to **appreciate** as **loathe** is to
   a. Detest.
   b. Hate.
   c. Appreciate.
   d. Dislike.
   e. Fear.

22. **Thanksgiving** is to **November** as
   a. Summer is to vacation.
   b. Easter is to spring.
   c. Labor Day is to January.
   d. Holiday is to celebration.
   e. Christmas is to December.

23. **Alligator** is to **reptile** as **elephant** is to
   a. Mammal.
   b. Animal.
   c. Australia.
   d. Marsupial.
   e. Bear.

24. **Nylon** is to **parachute** as
   a. Plexiglass is to glass.
   b. Neoprene is to wetsuit.
   c. Sweater is to wool.
   d. Wood is to bark.
   e. Ice is to cube.

25. **Painter** is to **easel** as **weaver** is to
   a. Pattern.
   b. Tapestry.
   c. Yarn.
   d. Needle.
   e. Loom.

Answer Explanations

1. C: This is a part/whole analogy. A chapter a section, or portion, of a book. All of the choices relate to literary topics, but the best option is paragraph is to essay. A paragraph is a section, or building block of an essay in much the same way that a chapter is in a book. The word pairs in Choices *A, B,* and *E* are best described as near synonyms, but not necessarily parts of one another. Choice *D,* dialogue is to play, does include more of a part to the whole relationship, but dialogue is the way the story is conveyed in a play (like sentences in a book). A better matching analogy to chapter is to book would be scene or act is to play, since plays are divided into scenes or larger acts.

2. A: This is a type of category analogy. A dress is a type of garment. Garment is the broad category, and dress is the specific example used. Diesel is a type of fuel, so it holds the same relationship. Month is not a type of year; it's part of the year. A suit is not a type of tie, but it might be worn with a tie. Clothing is not a type of wardrobe; it is stored in a wardrobe. Lastly, a coat isn't a type of winter; it is a garment worn in the winter.

3. E: This is a provider/provision analogy. The analogy focuses on where the mode of transportation is stored or housed when not in use. A garage is where a car is sheltered and kept when not in use, much like a hanger for an airplane. Choice *C,* airport, might be an appealing choice, but an airport doesn't have as precise of a relationship to a plane as does a garage to a car. An airport might be where planes are located before use or where one might see a lot of planes, like a parking lot for a car.

4. C: This is an intensity analogy, or possibly an antonym analogy. The prevailing connection is somewhat opposite meanings, or certainly opposite intensities. Fluid that is trickling is barely moving or of low volume, while gushing fluid is moving fast and often in a larger volume. The only choices that are related by intensity or are antonyms are Choices *C* and *D,* but *D* reverses the relationship. Sob is a more significant cry versus a small sniffle. Tepid is lukewarm or slightly warm, while scorching is very hot.

5. B: This is a provider/provision analogy. The connection is a condition treated by a certain medical professional. Acne may be treated by a dermatologist, a skin doctor. A cataract, an eye condition, is treated by an ophthalmologist. A psychologist treats psychological or mental conditions. An otolaryngologist is an ear, nose, and throat doctor, so one would not treat a cataract. Similarly, an otologist is an ear doctor, and an orthopedist treats skeletal issues.

6. C: This is a cause and effect analogy. The common thread is a physical reaction (effect) to a cause. Someone might jump from a surprise. Choice *C,* maintains this relationship because someone might chuckle in response to a joke. Choice *E* does contain a type of physical reaction effect (crying) and a cause (sadness), but the relationship is reversed. Moreover, "sadness" isn't really an event or occurrence the same way that a surprise or joke is. Sadness is an emotion.

7. B: This is tool/use analogy. The connection is a simple action that the tool (in this case, a utensil) is used for. The question is made easier by providing the next tool (utensil). A knife is used to slice things (bread, apple, etc.). A fork can be used to spear pieces of food, so the pieces can be lifted to the mouth for eating. Therefore, Choice *B* is the best option. A fork is not used for cutting; that is the function of a knife. Therefore, Choice *A* is incorrect. Choices *C* and *D* are incorrect because they aren't actions that the fork is used for. Choice *E,* eat, is technically an action the fork is used for, although it is not as precise and similar to spear, making Choice *B* a better answer.

8. D: This is a grammatical analogy. It is comparing a verb in the simple present tense with the same verb in the simple past tense. *Ate* is the past tense of *eat*, and *spun* is the past tense of *spin*.

9. D: This is a synonyms analogy, which matches terms based on their similar meanings. *Obviate* is a verb that can mean to prevent or avoid, or to render something unnecessary. It is usually used with an object, one that is often a difficulty or disadvantage. For example, wearing a helmet while cycling can obviate the risk of a skull injury should a fall occur. *Preclude* is also a verb. It means to prevent something from occurring or existing. For example, a thunderstorm may preclude a picnic. It can also mean to exclude from something. For example, an inability to get wet after surgery would preclude the patient from swimming. Of the answer choices provided, Choice *D*, appease and placate, are the only other synonyms. Like the terms in the prompt, these words are both verbs. *Placate* means to pacify, calm, or make a person less angry or upset. Similarly, *appease* means to pacify or calm someone by fulfilling their demands.

10. A: This can be considered a type of characteristic analogy because it is matching a unit of measure with an example of what that unit measures. An *acre* is an example of a unit of measure for area. Plots of land can be measured in acres, and this value gives information about how much area of land that plot occupies. A *fathom* is an example of a unit to measure depth. It is typically used to refer to water depth.

11. A: This is a characteristic analogy. The connection lies in what color results when yellow is added to the primary color. Green is the secondary color that is formed when yellow is added to blue, just as orange is the secondary color created when yellow is added to red.

12. D: This is a parts/whole analogy. The connection is the outer removable, inedible layer of the fruit. The best choice is *D*, shell is to coconut, because the shell is also the outer, inedible part of the coconut. While peaches do have fuzz, the fuzz is an edible portion of the skin. Therefore, Choice *E* is not as closely related to the question stem as the pair in Choice *D*.

13. E: This is a provider/provision analogy. The connection is a type of person or quality of a person and what that personality trait provides (the output, as a noun). A *sycophant* is someone who dishes out a lot of flattery, or insincere praise, often to better his or her situation. A *raconteur* dishes out or regales people with anecdotes and stories in an interesting way.

14. E: This is an antonyms analogy that matches adjectives with opposite meanings. A *viscous* fluid is thick and slowly moving, while a runny one is thin and flows freely. The only answer choice that is also a pair of adjectives that are antonyms is Choice *E*. When used as an adjective, *obscure* refers to something inconspicuous, unnoticeable, or ambiguous. It usually is used to refer to something with an unclear meaning or hard to understand, such as obscure intentions or the use of obscure language. *Unequivocal* is an adjective that means essentially the exact opposite—unambiguous or leaving no doubt.

15. B: This is another antonyms analogy. *Zenith* and *nadir* are astronomical nouns with opposite meanings that have also been adopted into conversational (non-technical) English. In an astronomical sense, the *zenith* is the point in the celestial sphere or sky that lies directly above the observer, while the *nadir* is the point directly below the observer. These terms have been incorporated into common language to mean the very top or culminating point of something (the *zenith*), and the very bottom, lowest, or worst (the *nadir*). For example, the *zenith* of triathlete's athletic career might be winning the Ironman World Championships, while the *nadir* might be crashing his or her bike during a race and fracturing a bone. In this question, the term *valley* is provided for the next pair. Like *nadir*, a *valley* is a very low point. The correct answer will then be a high point, which is best captured by Choice *B*, pinnacle.

16. E: This analogy matches a characteristic or type of person with the quality that type of person possesses. A *pundit* is an expert in a certain field or particular subject. A pundit has *expertise*. A scholar has learned or gained knowledge in the areas he or she has studied. They have *erudition*.

17. E: This analogy matches an occupation with what someone in that job creates and uses as a plan for their work. An architect creates a blueprint to be a rendering of the plan for the structures that builders will use to erect the building, much like a composer creates a score that musicians will follow to play the music. The other answer choices do not maintain this same relationship.

18. C: This is a type of tool/user analogy. Rather than simply being a "tool" used by the user, it pairs the user with what they seek. A detective searches for evidence or clues, while a prospector searches for gold.

19. B: This is a type of tool/use analogy. It matches a tool and what it is used to measure. An odometer is used to measure distance traveled in a car. A caliper is tool used to measure thickness. For example, personal trainers use skinfold calipers to measure the thickness of various folds of skin to estimate body fat.

20. A: This is a category analogy. The connection is the way that food is classified. A radish is a type of vegetable and a garbanzo bean (chickpea) is a type of legume. Pineapples are not a type of berry, lettuce is not a type of spinach, and cucumber is not a type of salad. Therefore, Choices *B, C,* and *D* are incorrect. Choice *E* may look appealing because citrus is a type of fruit, but "citrus" isn't a specific fruit. For the analogy to hold, it would need to be a specific citrus fruit, like lemon.

21. D: This is an intensity analogy. *Adore* is a stronger version of *appreciate*. *Loathe* is a stronger version of *dislike*. *Detest* and *hate,* Choices *A* and *B,* are more synonymous with *loathe* (rather than a different intensity), so they are not the best choice. Choice *C* is an antonym, and Choice *E* is unrelated.

22. E: This analogy makes use of the temporal relationship between an event (a holiday) and the calendar month that it occurs. Thanksgiving is a holiday in November, like Christmas is a holiday in December. Choice *C* is incorrect because Labor Day is in September. The other choices do not relate a specific holiday to the particular month in which they occur.

23. A: This is a category analogy. The common thread is the classification of the animal. An alligator is a type of reptile, just as an elephant is a type of mammal.

24. B: This is a source/comprised of analogy that focuses on pairing a raw material with an item that's created from it. Nylon is used to make parachutes just as neoprene is used to make wetsuits. Although sweaters are made from wool, the relationship is reversed. Test takers should remember to be careful about maintaining the same order of the words in the relationship when solving analogies; otherwise, the meaning is changed.

25. E: This is a tool/user analogy. The connection is the type of tool the artist uses to hold and create their work. An easel holds the paper or canvas that a painter uses to create a painting much like a loom is the apparatus used to hold and weave yarns into a blanket or other tapestry.

Arithmetic Reasoning

Numbers and Algebra

Definitions

Whole numbers are the numbers 0, 1, 2, 3, ... Examples of other whole numbers would be 413 and 8,431. Notice that numbers such as 4.13 and $\frac{1}{4}$ are not included in whole numbers. **Counting numbers**, also known as **natural numbers**, consist of all whole numbers except for the zero. In set notation, the natural numbers are the set $\{1, 2, 3, ...\}$. The entire set of whole numbers and negative versions of those same numbers comprise the set of numbers known as **integers.** Therefore, in set notation, the integers are $\{..., -3, -2, -1, 0, 1, 2, 3, ...\}$. Examples of other integers are −4,981 and 90,131. A number line is a great way to visualize the integers. Integers are labeled on the following number line:

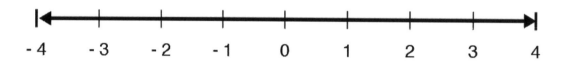

The arrows on the right- and left-hand sides of the number line show that the line continues indefinitely in both directions.

Fractions also exist on the number line as parts of a whole. For example, if an entire pie is cut into two pieces, each piece is half of the pie, or $\frac{1}{2}$. The top number in any fraction, known as the **numerator,** defines how many parts there are. The bottom number, known as the **denominator,** states how many pieces the whole is divided into. Fractions can also be negative or written in their corresponding decimal form.

A **decimal** is a number that uses a decimal point and numbers to the right of the decimal point representing the part of the number that is less than 1. For example, 3.5 is a decimal and is equivalent to the fraction $\frac{7}{2}$ or mixed number $3\frac{1}{2}$. The decimal is found by dividing 2 into 7. Other examples of fractions are $\frac{2}{7}, \frac{-3}{14}$, and $\frac{14}{27}$.

Any number that can be expressed as a fraction is known as a **rational number.** Basically, if a and b are any integers and $b \neq 0$, then $\frac{a}{b}$ is a rational number. Any integer can be written as a fraction where the denominator is 1, so therefore the rational numbers consist of all fractions and all integers.

Any number that is not rational is known as an **irrational number.** Consider the number $\pi = 3.141592654 ....$ The decimal portion of that number extends indefinitely. In that situation, a number can never be written as a fraction. Another example of an irrational number is $\sqrt{2} = 1.414213662 ....$ Again, this number cannot be written as a ratio of two integers.

Together, the set of all rational and irrational numbers makes up the **real numbers.** The number line contains all real numbers. To graph a number other than an integer on a number line, it needs to be plotted between two integers. For example, 3.5 would be plotted halfway between 3 and 4.

Even numbers are integers that are divisible by 2. For example, 6, 100, 0, and −200 are all even numbers. **Odd numbers** are integers that are not divisible by 2. If an odd number is divided by 2, the result is a fraction. For example, −5, 11, and −121 are odd numbers.

Prime numbers consist of natural numbers greater than 1 that are not divisible by any other natural numbers other than themselves and 1. For example, 3, 5, and 7 are prime numbers. If a natural number is not prime, it is known as a **composite number**. 8 is a composite number because it is divisible by both 2 and 4, which are natural numbers other than itself and 1.

The **absolute value** of any real number is the distance from that number to 0 on the number line. The absolute value of a number can never be negative. For example, the absolute value of both 8 and −8 is 8 because they are both 8 units away from 0 on the number line. This is written as $|8| = |-8| = 8$.

Writing Numbers Using Base-10 Numerals, Number Names, and Expanded Form

The **base-10 number system** is also called the **decimal system of naming numbers**. There is a decimal point that sets the value of numbers based on their position relative to the decimal point. The order from the decimal point to the right is the tenths place, then hundredths place, then thousandths place. Moving to the left from the decimal point, the place value is ones, tens, hundreds, etc. The number 2,356 can be described in words as "two thousand three hundred fifty-six." In expanded form, it can be written as $(2 \times 1,000) + (3 \times 100) + (5 \times 10) + (6 \times 1)$. The expanded form shows the value each number holds in its place. The number 3,093 can be written in words as "three thousand ninety-three." In expanded form, it can be expressed as $(3 \times 1,000) + (0 \times 100) + (9 \times 10) + (3 \times 1)$. Notice that the zero is added in the expanded form as a place holder. There are no hundreds in the number, so a zero is written in the hundreds place.

Composing and Decomposing Multidigit Numbers

Composing and decomposing numbers reveals the place value held by each number 0 through 9 in each position. For example, the number 17 is read as "seventeen." It can be decomposed into the numbers 10 and 7. It can be described as 1 group of ten and 7 ones. The one in the tens place represents one set of ten. The seven in the ones place represents seven sets of one. Added together, they make a total of seventeen. The number 48 can be written in words as "forty-eight." It can be decomposed into the numbers 40 and 8, where there are 4 groups of ten and 8 groups of one. The number 296 can be decomposed into 2 groups of one hundred, 9 groups of ten, and 6 groups of one. There are two hundreds, nine tens, and six ones. Decomposing and composing numbers lays the foundation for visually picturing the number and its place value, and adding and subtracting multiple numbers with ease.

The Place and Value of a Digit

Each number in the base-10 system is made of the numbers 0–9, located in different places relative to the decimal point. Based on where the numbers fall, the value of a digit changes. For example, the number 7,509 has a seven in the thousands place. This means there are seven groups of one thousand. The number 457 has a seven in the ones place. This means there are seven groups of one. Even though there is a seven in both numbers, the place of the seven tells the value of the digit. A practice question may ask the place and value of the 4 in 3,948. The four is found in the tens place, which means four represents the number 40, or four groups of ten. Another place value may be on the opposite side of the decimal point. A question may ask the place and value of the 8 in the number 203.80. In this case, the eight is in the tenths place because it is in the first place to the right of the decimal point. It holds a value of eight-tenths, or eight groups of one-tenth.

The Relative Value of a Digit

The value of a digit is found by recognizing its place relative to the rest of the number. For example, the number 569.23 contains a 6. The position of the 6 is two places to the left of the decimal, putting it in the tens place. The tens place gives it a value of 60, or six groups of ten. The number 39.674 has a 4 in it. The number 4 is located three places to the right of the decimal point, placing it in the thousandths place. The value of the 4 is four-thousandths, because of its position relative to the other numbers and to the decimal. It can be described as 0.004 by itself, or four groups of one-thousandths. The numbers 100 and 0.1 are both made up of ones and zeros. The first number, 100, has a 1 in the hundreds place, giving it a value of one hundred. The second number, 0.1, has a 1 in the tenths place, giving that 1 a value of one-tenth. The place of the number gives it the value.

Rounding Multidigit Numbers

Numbers can be rounded by recognizing the place value where the rounding takes place, then looking at the number to the right. If the number to the right is five or greater, the number to be rounded goes up one. If the number to the right is four or less, the number to be rounded stays the same. For example, the number 438 can be rounded to the tens place. The number 3 is in the tens place and the number to the right is 8. Because the 8 is 5 or greater, the 3 then rounds up to a 4. The rounded number is 440. Another number, 1,394, can be rounded to the thousands place. The number in the thousands place is 1, and the number to the right is 3. As the 3 is 4 or less, it means the 1 stays the same and the rounded number is 1,000. Rounding is also a form of estimating. The number 9.58 can be rounded to the tenths place. The number 5 is in the tenths place, and the number 8 is to the right of it. because 8 is 5 or greater, the 5 changes to a 6. The rounded number becomes 9.6.

Prime and Composite Numbers

A **prime number** is a whole number greater than 1 that can only be divided by 1 and itself. Examples are 2, 3, 5, 7, and 11. A **composite number** can be evenly divided by a number other than 1 and itself. Examples of composite numbers are 4 and 9. Four can be divided evenly by 1, 2, and 4. Nine can be divided evenly by 1, 3, and 9. When given a list of numbers, one way to determine which ones are prime or composite is to find the **prime factorization** of each number. For example, a list of numbers may include 13 and 24. The prime factorization of 13 is 1 and 13 because those are the only numbers that go into it evenly, so it is a prime number. The prime factorization of 24 is $2 \times 2 \times 2 \times 3$ because those are the prime numbers that multiply together to get 24. This also shows that 24 is a composite number because 2 and 3 are factors along with 1 and 24.

Multiples of Numbers

A **multiple** of a number is the result of multiplying that number by an integer. For example, some multiples of 3 are 6, 9, 12, 15, and 18. These multiples are found by multiplying 3 by 2, 3, 4, 5, and 6, respectively. Some multiples of 5 include 5, 10, 15, and 20. This also means that 5 is a factor of 5, 10, 15, and 20. Some questions may ask which numbers in a list are multiples of a given number. For example, find and circle the multiples of 12 in the following list: 136, 144, 312, 400. If a number is evenly divisible by 12, then it is a multiple of 12. The numbers 144 and 312 are multiples of 12 because 12 times 12 is 144, and 12 times 26 is 312. The other numbers, 136 and 400, are not multiples because they yield a number with a fractional component when divided by 12.

Factorization

Factorization is the process of breaking up a mathematical quantity, such as a number or polynomial, into a product of two or more factors. For example, a factorization of the number 16 is $16 = 8 \times 2$. If multiplied out, the factorization results in the original number. A **prime factorization** is a specific factorization when the number is factored completely using prime numbers only. For example, the prime factorization of 16 is $16 = 2 \times 2 \times 2 \times 2$. A factor tree can be used to find the prime factorization of any number. Within a factor tree, pairs of factors are found until no other factors can be used, as in the following factor tree of the number 84:

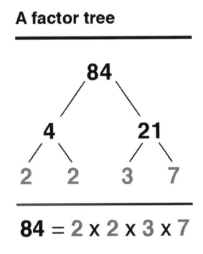

A factor tree

$$84 = 2 \times 2 \times 3 \times 7$$

It first breaks 84 into 21×4, which is not a prime factorization. Then, both 21 and 4 are factored into their primes. The final numbers on each branch consist of the numbers within the prime factorization. Therefore, $84 = 2 \times 2 \times 3 \times 7$. Factorization can be helpful in finding greatest common divisors and least common denominators.

Also, a factorization of an algebraic expression can be found. Throughout the process, a more complicated expression can be decomposed into products of simpler expressions. To factor a polynomial, first determine if there is a greatest common factor. If there is, factor it out. For example, $2x^2 + 8x$ has a greatest common factor of $2x$ and can be written as $2x(x + 4)$. Once the greatest common monomial factor is factored out, if applicable, count the number of terms in the polynomial. If there are two terms, is it a difference of squares, a sum of cubes, or a difference of cubes?

If so, the following rules can be used:

Difference of Squares:

$$a^2 - b^2 = (a + b)(a - b)$$

Sum of Cubes:

$$a^3 + b^3 = (a + b)(a^2 - ab + b^2)$$

Difference of Cubes

$$a^3 - b^3 = (a - b)(a^2 + ab + b^2)$$

If there are three terms, and if the trinomial is a perfect square trinomial, it can be factored into the following:

$$a^2 + 2ab + b^2 = (a + b)^2$$

$$a^2 - 2ab + b^2 = (a - b)^2$$

If not, try factoring into a product of two binomials by trial and error into a form of $(x + p)(x + q)$. For example, to factor $x^2 + 6x + 8$, determine what two numbers have a product of 8 and a sum of 6. Those numbers are 4 and 2, so the trinomial factors into $(x + 2)(x + 4)$.

Finally, if there are four terms, try factoring by grouping. First, group terms together that have a common monomial factor. Then, factor out the common monomial factor from the first two terms. Next, look to see if a common factor can be factored out of the second set of two terms that results in a common binomial factor. Finally, factor out the common binomial factor of each expression, for example:

$$xy - x + 5y - 5 = x(y - 1) + 5(y - 1) = (y - 1)(x + 5)$$

After the expression is completely factored, check to see if the factorization is correct by multiplying to try to obtain the original expression. Factorizations are helpful in solving equations that consist of a polynomial set equal to 0. If the product of two algebraic expressions equals 0, then at least one of the factors is equal to 0. Therefore, factor the polynomial within the equation, set each factor equal to 0, and solve. For example, $x^2 + 7x - 18 = 0$ can be solved by factoring into $(x + 9)(x - 2) = 0$. Set each factor equal to 0, and solve to obtain $x = -9$ and $x = 2$.

Converting Non-Negative Fractions, Decimals, and Percentages

Within the number system, different forms of numbers can be used. It is important to be able to recognize each type, as well as work with, and convert between, the given forms. The **real number system** comprises natural numbers, whole numbers, integers, rational numbers, and irrational numbers. Natural numbers, whole numbers, integers, and irrational numbers typically are not represented as fractions, decimals, or percentages. Rational numbers, however, can be represented as any of these three forms. A **rational number** is a number that can be written in the form $\frac{a}{b}$, where a and b are integers, and b is not equal to zero. In other words, rational numbers can be written in a fraction form. The value a is the **numerator,** and b is the **denominator.** If the numerator is equal to zero, the entire fraction is equal to zero. Non-negative fractions can be less than 1, equal to 1, or greater than 1. Fractions are less than 1 if the numerator is smaller (less than) than the denominator. For example, $\frac{3}{4}$ is less than 1. A fraction is equal to 1 if the numerator is equal to the denominator. For instance, $\frac{4}{4}$ is equal to 1. Finally, a fraction is greater than 1 if the numerator is greater than the denominator: the fraction $\frac{11}{4}$ is greater than 1. When the numerator is greater than the denominator, the fraction is called an **improper fraction**. An improper fraction can be converted to a **mixed number**, a combination of both a whole number and a fraction. To convert an improper fraction to a mixed number, divide the numerator by the denominator. Write down the whole number portion, and then write any remainder over the original denominator. For example, $\frac{11}{4}$ is equivalent to $2\frac{3}{4}$. Conversely, a mixed number can be converted to an improper fraction by multiplying the denominator by the whole number and adding that result to the numerator.

Fractions can be converted to decimals. With a calculator, a fraction is converted to a decimal by dividing the numerator by the denominator. For example, $\frac{2}{5} = 2 \div 5 = 0.4$. Sometimes, rounding might be necessary. Consider $\frac{2}{7} = 2 \div 7 = 0.28571429$. This decimal could be rounded for ease of use, and if it needed to be rounded to the nearest thousandth, the result would be 0.286. If a calculator is not available, a fraction can be converted to a decimal manually. First, find a number that, when multiplied by the denominator, has a value equal to 10, 100, 1,000, etc. Then, multiply both the numerator and denominator times that number. The decimal form of the fraction is equal to the new numerator with a decimal point placed as many place values to the left as there are zeros in the denominator. For example, to convert $\frac{3}{5}$ to a decimal, multiply both the numerator and denominator times 2, which results in $\frac{6}{10}$. The decimal is equal to 0.6 because there is one zero in the denominator, and so the decimal place in the numerator is moved one unit to the left. In the case where rounding would be necessary while working without a calculator, an approximation must be found. A number close to 10, 100, 1,000, etc. can be used. For example, to convert $\frac{1}{3}$ to a decimal, the numerator and denominator can be multiplied by 33 to turn the denominator into approximately 100, which makes for an easier conversion to the equivalent decimal. This process results in $\frac{33}{99}$ and an approximate decimal of 0.33. Once in decimal form, the number can be converted to a percentage. Multiply the decimal by 100 and then place a percent sign after the number. For example, .614 is equal to 61.4%. In other words, move the decimal place two units to the right and add the percent symbol.

Identifying Integers

Integers include zero, and both positive and negative numbers with no fractional component. Examples of integers are −3, 5, 120, −47, and 0. Numbers that are not integers include 1.3333, ½, −5.7, and 4½. Integers can be used to describe different real-world situations. If a scuba diver were to dive 50 feet down into the ocean, his position can be described as −50, in relation to sea level. If while traveling in Denver, Colorado, a car has an elevation reading of 2300 feet, the integer 2300 can be used to describe the feet above sea level. Integers can be used in many different ways to describe situations with whole numbers and zero.

Integers can also be added and subtracted as situations change. If the temperature in the morning is 45 degrees, and it dropped to 33 degrees in the afternoon, the difference can be found by subtracting the integers 45 and 33 to get a change of 12 degrees. If a submarine was at a depth of 100 feet below sea level, then rose 35 feet, the new depth can be found by adding −100 to 35. The following equation can be used to model the situation with integers: −100 + 35 −65. The answer of −65 reveals the new depth of the submarine, 65 feet below sea level.

Arithmetic Operations with Rational Numbers

The four basic operations include addition, subtraction, multiplication, and division. The result of addition is a sum, the result of subtraction is a difference, the result of multiplication is a product, and the result of division is a quotient. Each type of operation can be used when working with rational numbers; however, the basic operations need to be understood first while using simpler numbers before working with fractions and decimals.

Performing these operations should first be learned using whole numbers. Addition needs to be done column by column. To add two whole numbers, add the ones column first, then the tens columns, then the hundreds, etc. If the sum of any column is greater than 9, a one must be carried over to the next column. For example, the following is the result of 482 + 924:

$$\begin{array}{r} \overset{1}{} \\ 482 \\ +924 \\ \hline 1406 \end{array}$$

Notice that the sum of the tens column was 10, so a one was carried over to the hundreds column. Subtraction is also performed column by column. Subtraction is performed in the ones column first, then the tens, etc. If the number on top is smaller than the number below, a one must be borrowed from the column to the left. For example, the following is the result of 5,424 – 756:

$$\begin{array}{r} 4\ 13\ 11\ 14 \\ \cancel{5}\ \cancel{4}\ \cancel{2}\ \cancel{4} \\ -\ 7\ 5\ 6 \\ \hline 4\ 6\ 6\ 8 \end{array}$$

Notice that a one is borrowed from the tens, hundreds, and thousands place. After subtraction, the answer can be checked through addition. A check of this problem would be to show that 756 + 4,668 = 5,424.

Multiplication of two whole numbers is performed by writing one on top of the other. The number on top is known as the **multiplicand,** and the number below is the **multiplier.** Perform the multiplication by multiplying the multiplicand by each digit of the multiplier. Make sure to place the ones value of each result under the multiplying digit in the multiplier. Each value to the right is then a 0. The product is found by adding each product. For example, the following is the process of multiplying 46 times 37 where 46 is the multiplicand and 37 is the multiplier:

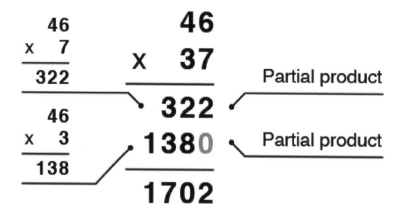

Finally, division can be performed using long division. When dividing a number by another number, the first number is known as the **dividend,** and the second is the **divisor.** For example, with $a \div b = c$, a is the dividend, *b* is the divisor, and *c* is the quotient. For long division, place the dividend within the division symbol and the divisor on the outside. For example, with 8,764 ÷ 4, refer to the first problem in the diagram below. First, there are 2 4's in the first digit, 8. This number 2 gets written above the 8. Then, multiply 4 times 2 to get 8, and that product goes below the 8. Subtract to get 8, and then carry down the 7. Continue the same steps. 7 ÷ 4 = 1 R3, so 1 is written above the 7. Multiply 4 times 1 to get 4, and write it below the 7. Subtract to get 3, and carry the 6 down next to the 3. Resulting steps give a 9 and a 1. The final subtraction results in a 0, which means that 8,764 is divisible by 4. There are no remaining numbers.

The second example shows that 4,536 ÷ 216 = 21. The steps are a little different because 216 cannot be contained in 4 or 5, so the first step is placing a 2 above the 3 because there are 2 216's in 453. Finally, the third example shows that 546 ÷ 31 = 17 R19. The 19 is a remainder. Notice that the final subtraction does not result in a 0, which means that 546 is not divisible by 31. The remainder can also be written as a fraction over the divisor to say that $546 \div 31 = 17\frac{19}{31}$.

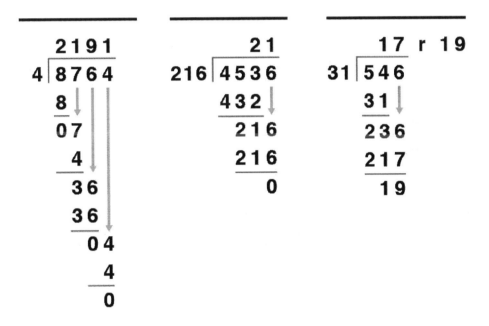

If a division problem relates to a real-world application, and a remainder does exist, it can have meaning. For example, consider the third example, 546 ÷ 31 = 17R19. Let's say that we had $546 to spend on calculators that cost $31 each, and we wanted to know how many we could buy. The division problem would answer this question. The result states that 17 calculators could be purchased, with $19 left over. Notice that the remainder will never be greater than or equal to the divisor.

Once the operations are understood with whole numbers, they can be used with integers. There are many rules surrounding operations with negative numbers. First, consider addition with integers. The sum of two numbers can first be shown using a number line. For example, to add −5 + (−6), plot the point −5 on the number line. Then, because a negative number is being added, move 6 units to the left. This process results in landing on −11 on the number line, which is the sum of −5 and −6. If adding a positive number, move to the right. Visualizing this process using a number line is useful for understanding; however, it is not efficient. A quicker process is to learn the rules. When adding two numbers with the same sign, add the absolute values of both numbers, and use the common sign of both numbers as the sign of the sum. For example, to add $-5 + (-6)$, add their absolute values $5 + 6 = 11$. Then, introduce a negative number because both addends are negative. The result is -11. To add two integers with unlike signs, subtract the lesser absolute value from the greater absolute value, and apply the sign of the number with the greater absolute value to the result. For example, the sum $-7 + 4$ can be computed by finding the difference $7 - 4 = 3$ and then applying a negative because the value with the larger absolute value is negative. The result is -3. Similarly, the sum $-4 + 7$ can be found by computing the same difference but leaving it as a positive result because the addend with the larger absolute value is positive. Also, recall that any number plus 0 equals that number. This is known as the **Addition Property of 0.**

Subtracting two integers can be computed by changing to addition to avoid confusion. The rule is to add the first number to the opposite of the second number. The opposite of a number is the number on the other side of 0 on the number line, which is the same number of units away from 0. For example, -2 and 2 are opposites. Consider $4 - 8$. Change this to adding the opposite as follows: $4 + (-8)$. Then, follow the rules of addition of integers to obtain -4. Secondly, consider $-8 - (-2)$. Change this problem to adding the opposite as $-8 + 2$, which equals -6. Notice that subtracting a negative number is really adding a positive number.

Multiplication and division of integers are actually less confusing than addition and subtraction because the rules are simpler to understand. If two factors in a multiplication problem have the same sign, the result is positive. If one factor is positive and one factor is negative, the result, known as the *product,* is negative. For example, $(-9)(-3) = 27$ and $9(-3) = -27$. Also, a number times 0 always results in 0. If a problem consists of more than a single multiplication, the result is negative if it contains an odd number of negative factors, and the result is positive if it contains an even number of negative factors. For example, $(-1)(-1)(-1)(-1) = 1$ and $(-1)(-1)(-1)(-1)(-1) = -1$. These two examples of multiplication also bring up another concept. Both are examples of repeated multiplication, which can be written in a more compact notation using exponents. The first example can be written as $(-1)^4 = 1$, and the second example can be written as $(-1)^5 = -1$. Both are exponential expressions, -1 is the base in both instances, and 4 and 5 are the respective exponents. Note that a negative number raised to an odd power is always negative, and a negative number raised to an even power is always positive. Also, $(-1)^4$ is not the same as -1^4. In the first expression, the negative is included in the parentheses, but it is not in the second expression. The second expression is found by evaluating 1^4 first to get 1 and then by applying the negative sign to obtain -1.

A similar theory applies within division. First, consider some vocabulary. When dividing 14 by 2, it can be written in the following ways: $14 \div 2 = 7$ or $\frac{14}{2} = 7$. 14 is the **dividend,** 2 is the **divisor,** and 7 is the **quotient.** If two numbers in a division problem have the same sign, the quotient is positive. If two numbers in a division problem have different signs, the quotient is negative. For example:

$$14 \div (-2) = -7, \text{ and } -14 \div (-2) = 7$$

To check division, multiply the quotient by the divisor to obtain the dividend. Also, remember that 0 divided by any number is equal to 0. However, any number divided by 0 is undefined. It just does not make sense to divide a number by 0 parts.

If more than one operation is to be completed in a problem, follow the **Order of Operations**. The mnemonic device, PEMDAS, for the order of operations states the order in which addition, subtraction, multiplication, and division needs to be done. It also includes when to evaluate operations within grouping symbols and when to incorporate exponents. PEMDAS, which some remember by thinking "please excuse my dear Aunt Sally," refers to parentheses, exponents, multiplication, division, addition, and subtraction. First, within an expression, complete any operation that is within parentheses, or any other grouping symbol like brackets, braces, or absolute value symbols. Note that this does not refer to the case when parentheses are used to represent multiplication like (2)(5). In this case, an operation is not within parentheses like it is in (2 × 5). Then, any exponents must be computed. Next, multiplication and division are performed from left to right.

Finally, addition and subtraction are performed from left to right. The following is an example in which the operations within the parentheses need to be performed first, so the order of operations must be applied to the exponent, subtraction, addition, and multiplication within the grouping symbol:

$$9-3(3^2-3+4\cdot3)$$

$$9-3(3^2-3+4\cdot3) \quad \text{Work within the parentheses first}$$

$$= 9-3(9-3+12)$$

$$= 9-3(18)$$

$$= 9-54$$

$$= -45$$

Once the rules for integers are understood, move on to learning how to perform operations with fractions and decimals. Recall that a rational number can be written as a fraction and can be converted to a decimal through division. If a rational number is negative, the rules for adding, subtracting, multiplying, and dividing integers must be used. If a rational number is in fraction form, performing addition, subtraction, multiplication, and division is more complicated than when working with integers. First, consider addition. To add two fractions having the same denominator, add the numerators and then reduce the fraction. When an answer is a fraction, it should always be in lowest terms. **Lowest terms** means that every common factor, other than 1, between the numerator and denominator is divided out. For example:

$$\frac{2}{8}+\frac{4}{8}=\frac{6}{8}=\frac{6\div2}{8\div2}=\frac{3}{4}$$

Both the numerator and denominator of $\frac{6}{8}$ have a common factor of 2, so 2 is divided out of each number to put the fraction in lowest terms. If denominators are different in an addition problem, the fractions must be converted to have common denominators. The **least common denominator (LCD)** of all the given denominators must be found, and this value is equal to the **least common multiple (LCM)** of the denominators. This non-zero value is the smallest number that is a multiple of both denominators. Then, rewrite each original fraction as an equivalent fraction using the new denominator. Once in this form, apply the process of adding with like denominators. For example, consider $\frac{1}{3}+\frac{4}{9}$. The LCD is 9 because it is the smallest multiple of both 3 and 9. The fraction $\frac{1}{3}$ must be rewritten with 9 as its denominator. Therefore, multiply both the numerator and denominator by 3. Multiplying by $\frac{3}{3}$ is the same as multiplying by 1, which does not change the value of the fraction. Therefore, an equivalent fraction is $\frac{3}{9}$, and $\frac{1}{3}+\frac{4}{9}=$

$\frac{3}{9} + \frac{4}{9} = \frac{7}{9}$, which is in lowest terms. Subtraction is performed in a similar manner; once the denominators are equal, the numerators are then subtracted. The following is an example of addition of a positive and a negative fraction:

$$-\frac{5}{12} + \frac{5}{9} = -\frac{5 \times 3}{12 \times 3} + \frac{5 \times 4}{9 \times 4} = -\frac{15}{36} + \frac{20}{36} = \frac{5}{36}$$

Common denominators are not used in multiplication and division. To multiply two fractions, multiply the numerators together and the denominators together. Then, write the result in lowest terms. For example:

$$\frac{2}{3} \times \frac{9}{4} = \frac{18}{12} = \frac{3}{2}$$

Alternatively, the fractions could be factored first to cancel out any common factors before performing the multiplication. For example:

$$\frac{2}{3} \times \frac{9}{4} = \frac{2}{3} \times \frac{3 \times 3}{2 \times 2} = \frac{3}{2}$$

This second approach is helpful when working with larger numbers, as common factors might not be obvious. Multiplication and division of fractions are related because the division of two fractions is changed into a multiplication problem. Division of a fraction is equivalent to multiplication of the **reciprocal** of the second fraction, so that second fraction must be inverted, or "flipped," to be in reciprocal form. For example:

$$\frac{11}{15} \div \frac{3}{5} = \frac{11}{15} \times \frac{5}{3} = \frac{55}{45} = \frac{11}{9}$$

The fraction $\frac{5}{3}$ is the reciprocal of $\frac{3}{5}$. It is possible to multiply and divide numbers containing a mix of integers and fractions. In this case, convert the integer to a fraction by placing it over a denominator of 1. For example, a division problem involving an integer and a fraction is:

$$3 \div \frac{1}{2} = \frac{3}{1} \times \frac{2}{1} = \frac{6}{1} = 6$$

Finally, when performing operations with rational numbers that are negative, the same rules apply as when performing operations with integers. For example, a negative fraction times a negative fraction results in a positive value, and a negative fraction subtracted from a negative fraction results in a negative value.

Operations can be performed on rational numbers in decimal form. Recall that to write a fraction as an equivalent decimal expression, divide the numerator by the denominator. For example:

$$\frac{1}{8} = 1 \div 8 = 0.125$$

With the case of decimals, it is important to keep track of place value. To add decimals, make sure the decimal places are in alignment so that the numbers are lined up with their decimal points and add vertically. If the numbers do not line up because there are extra or missing place values in one of the numbers, then zeros may be used as placeholders. For example, 0.123 + 0.23 becomes:

$$
\begin{array}{r}
0.123 \\
+\ 0.230 \\
\hline
0.353
\end{array}
$$

Subtraction is done the same way. Multiplication and division are more complicated. To multiply two decimals, place one on top of the other as in a regular multiplication process and do not worry about lining up the decimal points. Then, multiply as with whole numbers, ignoring the decimals. Finally, in the solution, insert the decimal point as many places to the left as there are total decimal values in the original problem. Here is an example of a decimal multiplication:

$$
\begin{array}{rl}
0.52 & \textit{2 decimal places} \\
\times\quad 0.2 & \textit{1 decimal place} \\
\hline
0.104 & \textit{3 decimal places}
\end{array}
$$

The answer to 52 times 2 is 104, and because there are three decimal values in the problem, the decimal point is positioned three units to the left in the answer.

The decimal point plays an integral role throughout the whole problem when dividing with decimals. First, set up the problem in a long division format. If the divisor is not an integer, the decimal must be moved to the right as many units as needed to make it an integer. The decimal in the dividend must be moved to the right the same number of places to maintain equality. Then, division is completed normally. Here is an example of long division with decimals using the problem 12.72 ÷ 0.06:

First, move the decimal point in 12.72 and 0.06 two places to the right. This gives 1272 and 6.

**Long division
with decimals**

$$
\begin{array}{r}
212 \\
6\,\overline{)1272} \\
\underline{12} \\
07 \\
\underline{6} \\
12
\end{array}
$$

Because the decimal point is moved two units to the right in the divisor of 0.06 to turn it into the integer 6, it is also moved two units to the right in the dividend of 12.72 to make it 1,272. The result is 212. Now move the decimal place back to the left by two units to get 2.12. This is the answer to 12.72 ÷ 0.06. Also remember that a division problem can always be checked by multiplying the answer times the divisor to see if the result is equal to the dividend.

Sometimes it is helpful to round answers that are in decimal form. First, find the place to which the rounding needs to be done. Then, look at the digit to the right of it. If that digit is 4 or less, the number in the place value to its left stays the same, and everything to its right becomes a 0. This process is known as **rounding down**. If that digit is 5 or higher, round up by increasing the place value to its left by 1, and every number to its right becomes a 0. If those 0's are in decimals, they can be dropped. For example, 0.145 rounded to the nearest hundredth place would be rounded up to 0.15, and 0.145 rounded to the nearest tenth place would be rounded down to 0.1.

Another operation that can be performed on rational numbers is the square root. Dealing with real numbers only, the positive square root of a number is equal to one of the two repeated positive factors of that number. For example, $\sqrt{49} = \sqrt{7 \times 7} = 7$. A **perfect square** is a number that has a whole number as its square root. Examples of perfect squares are 1, 4, 9, 16, 25, etc. If a number is not a perfect square, an approximation can be used with a calculator. For example, $\sqrt{67} = 8.185$, rounded to the nearest thousandth place. The square root of a fraction involving perfect squares involves breaking up the problem into the square root of the numerator separate from the square root of the denominator. For example:

$$\sqrt{\frac{16}{25}} = \frac{\sqrt{16}}{\sqrt{25}} = \frac{4}{5}$$

If the fraction does not contain perfect squares, a calculator can be used. Therefore, $\sqrt{\frac{2}{5}} = 0.632$, rounded to the nearest thousandth place. A common application of square roots involves the Pythagorean Theorem. Given a right triangle, the sum of the squares of the two legs equals the square of the hypotenuse.

For example, consider the following right triangle:

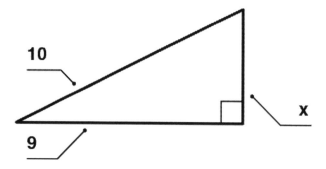

The missing side, x, can be found using the Pythagorean Theorem.

$$9^2 + x^2 = 10^2$$

$$81 + x^2 = 100$$

$$x^2 = 19$$

To solve for x, take the square root of both sides. Therefore, $x = \sqrt{19} = 4.36$, which has been rounded to two decimal places.

In addition to the square root, the cube root is another operation. If a number is a **perfect cube,** the cube root of that number is equal to one of the three repeated factors. For example:

$$\sqrt[3]{27} = \sqrt[3]{3 \times 3 \times 3} = 3$$

Also, unlike square roots, a negative number has a cube root. The result is a negative number. For example:

$$\sqrt[3]{-27} = \sqrt[3]{(-3)(-3)(-3)} = -3$$

Similar to square roots, if the number is not a perfect cube, a calculator can be used to find an approximation. Therefore, $\sqrt[3]{\frac{2}{3}} = 0.873$, rounded to the nearest thousandth place.

Higher-order roots also exist. The number relating to the root is known as the **index.** Given the following root, $\sqrt[3]{64}$, 3 is the index, and 64 is the **radicand.** The entire expression is known as the **radical.** Higher-order roots exist when the index is larger than 3. They can be broken up into two groups: even and odd roots. Even roots, when the index is an even number, follow the properties of square roots. A negative number does not have an even root, and an even root is found by finding the single factor that is repeated the same number of times as the index in the radicand. For example, the fifth root of 32 is equal to 2 because:

$$\sqrt[5]{32} = \sqrt[5]{2 \times 2 \times 2 \times 2 \times 2} = 2$$

Odd roots, when the index is an odd number, follow the properties of cube roots. A negative number has an odd root. Similarly, an odd root is found by finding the single factor that is repeated that many times to obtain the radicand. For example, the 4th root of 81 is equal to 3 because $3^4 = 81$. This radical is written as $\sqrt[4]{81} = 4$.

When performing operations in rational numbers, sometimes it might be helpful to round the numbers in the original problem to get a rough estimate of what the answer should be. For example, if you walked into a grocery store and had a $20 bill, your approach might be to round each item to the nearest dollar and add up all the items to make sure that you will have enough money when you check out. This process involves obtaining an estimation of what the exact total would be. In other situations, it might be helpful to round to the nearest $10 amount or $100 amount. **Front-end rounding** might be helpful as well in many situations. In this type of rounding, each number is rounded to the highest possible place value. Therefore, all digits except the first digit become 0. Consider a situation in which you are at the furniture store and want to estimate your total on three pieces of furniture that cost $434.99, $678.99, and $129.99. Front-end rounding would round these three amounts to $400, $700, and $100. Therefore, the estimate of your total would be $400 + $700 + $100 = $1,200, compared to the exact total of $1,243.97. In this situation, the estimate is not that far off the exact answer.

Rounding is useful in both approximating an answer when an exact answer is not needed and for comparison when an exact answer is needed. For instance, if you had a complicated set of operations to complete and your estimate was $1,000, if you obtained an exact answer of $100,000, something is off. You might want to check your work to see if a mistake was made because an estimate should not be that different from an exact answer. Estimates can also be helpful with square roots. If a square root of a number is not known, the closest perfect square can be found for an approximation. For example, $\sqrt{50}$ is not equal to a whole number, but 50 is close to 49, which is a perfect square, and $\sqrt{49} = 7$. Therefore, $\sqrt{50}$ is a little bit larger than 7. The actual approximation, rounded to the nearest thousandth, is 7.071.

Ordering and Comparing Rational Numbers

Ordering rational numbers is a way to compare two or more different numerical values. Determining whether two amounts are equal, less than, or greater than is the basis for comparing both positive and negative numbers. Also, a group of numbers can be compared by ordering them from the smallest amount to the largest amount. A few symbols are necessary to use when ordering rational numbers. The equals sign, =, shows that the two quantities on either side of the symbol have the same value. For example, $\frac{12}{3} = 4$ because both values are equivalent. Another symbol that is used to compare numbers is <, which represents "less than." With this symbol, the smaller number is placed on the left and the larger number is placed on the right. Always remember that the symbol's "mouth" opens up to the larger number. When comparing negative and positive numbers, it is important to remember that the number occurring to the left on the number line is always smaller and is placed to the left of the symbol. This idea might seem confusing because some values could appear at first glance to be larger, even though they are not. For example, $-5 < 4$ is read "negative 5 is less than 4." Here is an image of a number line for help:

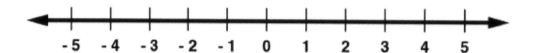

The symbol $\leq$ represents "less than or equal to," and it joins < with equality. Therefore, both $-5 \leq 4$ and $-5 \leq -5$ are true statements and "-5 is less than or equal to both 4 and -5." Other symbols are > and $\geq$, which represent "greater than" and "greater than or equal to." Both $4 \geq -1$ and $-1 \geq -1$ are correct ways to use these symbols.

Here is a chart of these four inequality symbols:

Symbol	Definition
<	less than
$\leq$	less than or equal to
>	greater than
$\geq$	greater than or equal to

Comparing integers is a straightforward process, especially when using the number line, but the comparison of decimals and fractions is not as obvious. When comparing two non-negative decimals, compare digit by digit, starting from the left. The larger value contains the first larger digit. For example, 0.1456 is larger than 0.1234 because the value 4 in the hundredths place in the first decimal is larger than the value 2 in the hundredths place in the second decimal. When comparing a fraction with a decimal, convert the fraction to a decimal and then compare in the same manner. Finally, there are a few options when comparing fractions. If two non-negative fractions have the same denominator, the fraction with the larger numerator is the larger value. If they have different denominators, they can be converted to equivalent fractions with a common denominator to be compared, or they can be converted to decimals to be compared. When comparing two negative decimals or fractions, a different approach must be used. It is important to remember that the smaller number exists to the left on the number line. Therefore, when comparing two negative decimals by place value, the number with the larger first place value is smaller due to the negative sign. Whichever value is closer to 0 is larger. For instance, -0.456 is larger than -0.498

because of the values in the hundredth places. If two negative fractions have the same denominator, the fraction with the larger numerator is smaller because of the negative sign.

Applying Estimation Strategies and Rounding Rules to Real-World Problems

Sometimes it is helpful to find an estimated answer to a problem rather than working out an exact answer. An estimation might be much quicker to find, and given the scenario, an estimation might be all that is required. For example, if Aria goes grocery shopping and has only a $100 bill to cover all of her purchases, it might be appropriate for her to estimate the total of the items she is purchasing to determine if she has enough money to cover them. Also, an estimation can help determine if an answer makes sense. For instance, if an answer in the 100s is expected, but the result is a fraction less than 1, something is probably wrong in the calculation.

The first type of estimation involves rounding. As mentioned, **rounding** consists of expressing a number in terms of the nearest decimal place like the tenth, hundredth, or thousandth place, or in terms of the nearest whole number unit like tens, hundreds, or thousands place. When rounding to a specific place value, look at the digit to the right of the place. If it is 5 or higher, round the number to its left up to the next value, and if it is 4 or lower, keep that number at the same value. For instance, 1,654.2674 rounded to the nearest thousand is 2,000, and the same number rounded to the nearest thousandth is 1,654.267. Rounding can be used in the scenario when grocery totals need to be estimated. Items can be rounded to the nearest dollar. For example, a can of corn that costs $0.79 can be rounded to $1.00, and then all other items can be rounded in a similar manner and added together. When working with larger numbers, it might make more sense to round to higher place values. For example, when estimating the total value of a dealership's car inventory, it would make sense to round the car values to the nearest thousands place. The price of a car that is on sale for $15,654 can be estimated at $16,000. All other cars on the lot could be rounded in the same manner, and then their sum can be found. Depending on the situation, it might make sense to calculate an over-estimate. For example, to make sure Aria has enough money at the grocery store, rounding up every time for each item would ensure that she will have enough money when it comes time to pay. A $.40 item rounded up to $1.00 would ensure that there is a dollar to cover that item. Traditional rounding rules would round $0.40 to $0, which does not make sense in this particular real-world setting. Aria might not have a dollar available at checkout to pay for that item if she uses traditional rounding. It is up to the customer to decide the best approach when estimating.

Estimating is also very helpful when working with measurements. Bryan is updating his kitchen and wants to retile the floor. Again, an over-measurement might be useful. Also, rounding to nearest half-unit might be helpful. For instance, one side of the kitchen might have an exact measurement of 14.32 feet, and the most useful measurement needed to buy tile could be estimating this quantity to be 14.5 feet. If the kitchen was rectangular and the other side measured 10.9 feet, Bryan might round the other side to 11 feet. Therefore, Bryan would find the total tile necessary according to the following area calculation: $14.5 \times 11 = 159.5$ square feet. To make sure he purchases enough tile, Bryan would probably want to purchase at least 160 square feet of tile. This is a scenario in which an estimation might be more useful than an exact calculation. Having more tile than necessary is better than having an exact amount, in case any tiles are broken or otherwise unusable.

Finally, estimation is helpful when exact answers are necessary. Consider a situation in which Sabina has many operations to perform on numbers with decimals, and she is allowed a calculator to find the result. Even though an exact result can be obtained with a calculator, there is always a possibility that Sabina could make an error while inputting the data. For example, she could miss a decimal place, or misuse a parenthesis, causing a problem with the actual order of operations. In this case, a quick estimation at the beginning would be helpful to make sure the final answer is given with the correct number of units. Sabina has to find the exact total of 10 cars listed for sale at the dealership. Each price has two decimal places included to account for both dollars and cents. If one car is listed at $21, 234.43 but Sabina incorrectly inputs into the calculator the price of $2,123.443, this error would throw off the final sum by almost $20,000. A quick estimation at the beginning, by rounding each price to the nearest thousands place and finding the sum of the prices, would give Sabina an amount to compare the exact amount to. This comparison would let Sabina see if an error was made in her exact calculation.

Percentages

Percentages are defined to be parts per one hundred. To convert a decimal to a percentage, move the decimal point two units to the right and place the percent sign after the number. Percentages appear in many scenarios in the real world. It is important to make sure the statement containing the percentage is translated to a correct mathematical expression. Be aware that it is extremely common to make a mistake when working with percentages within word problems.

An example of a word problem containing a percentage is the following: 35% of people speed when driving to work. In a group of 5,600 commuters, how many would be expected to speed on the way to their place of employment? The answer to this problem is found by finding 35% of 5,600. First, change the percentage to the decimal 0.35. Then compute the product: $0.35 \times 5,600 = 1,960$. Therefore, it would be expected that 1,960 of those commuters would speed on their way to work based on the data given. In this situation, the word "of" signals to use multiplication to find the answer.

Another way percentages are used is in the following problem: Teachers work 8 months out of the year. What percent of the year do they work? To answer this problem, find what percent of 12 the number 8 is, because there are 12 months in a year. Therefore, divide 8 by 12, and convert that number to a percentage: $\frac{8}{12} = \frac{2}{3} = 0.66\overline{6}$. The percentage rounded to the nearest tenth place tells us that teachers work 66.7% of the year. Percentages also appear in real-world application problems involving finding missing quantities like in the following question: 60% of what number is 75? To find the missing quantity, an equation can be used. Let x be equal to the missing quantity. Therefore, $0.60x = 75$. Divide each side by 0.60 to obtain 125. Therefore, 60% of 125 is equal to 75.

Sales tax is an important application relating to percentages because tax rates are usually given as percentages. For example, a city might have an 8% sales tax rate. Therefore, when an item is purchased with that tax rate, the real cost to the customer is 1.08 times the price in the store. For example, a $25 pair of jeans costs the customer $25 \times 1.08 = $27. Sales tax rates can also be determined if they are unknown when an item is purchased. If a customer visits a store and purchases an item for $21.44, but the price in the store was $19, they can find the tax rate by first subtracting $21.44 − $19 to obtain $2.44, the sales tax amount. The sales tax is a percentage of the in-store price. Therefore, the tax rate is $\frac{2.44}{19} = 0.128$, which has been rounded to the nearest thousandths place. In this scenario, the actual sales tax rate given as a percentage is 12.8%.

Proportions

Fractions appear in everyday situations, and in many scenarios, they appear in the real-world as ratios and in proportions. A **ratio** is formed when two different quantities are compared. For example, in a group of 50 people, if there are 33 females and 17 males, the ratio of females to males is 33 to 17. This expression can be written in the fraction form, $\frac{33}{17}$, or by using the ratio symbol, 33:17. The order of the number matters when forming ratios. In the same setting, the ratio of males to females is 17 to 33, which is equivalent to $\frac{17}{33}$ or 17:33. A **proportion** is an equation involving two ratios. The equation $\frac{a}{b} = \frac{c}{d}$, or $a:b = c:d$ is a proportion, for real numbers a, b, c, and d. Usually, in one ratio, one of the quantities is unknown, and cross-multiplication is used to solve for the unknown. Consider $\frac{1}{4} = \frac{x}{5}$. To solve for x, cross-multiply to obtain $5 = 4x$. Divide each side by 4 to obtain the solution $x = \frac{5}{4}$. It is also true that percentages are ratios in which the second term is 100. For example, 65% is 65:100 or $\frac{65}{100}$. Therefore, when working with percentages, one is also working with ratios.

Real-world problems frequently involve proportions. For example, consider the following problem: If 2 out of 50 pizzas are usually delivered late from a local Italian restaurant, how many would be late out of 235 pizzas? The following proportion would be solved with x as the unknown quantity of late pizza: $\frac{2}{50} = \frac{x}{235}$. Cross-multiplying results in $470 = 50x$. Divide both sides by 50 to obtain $x = \frac{470}{50}$, which in lowest terms is equal to $\frac{47}{5}$. In decimal form, this improper fraction is equal to 9.4. Because it does not make sense to answer this question with decimals (portions of pizza do not get delivered) the answer must be rounded. Traditional rounding rules would say that 9 pizzas would be expected to be delivered late. However, to be safe, rounding up to 10 pizzas out of 235 would probably make more sense.

Ratios and Rates of Change

Recall that a **ratio** is the comparison of two different quantities. Comparing 2 apples to 3 oranges results in the ratio 2:3, which can be expressed as the fraction $\frac{2}{3}$. Note that order is important when discussing ratios. The number mentioned first is the numerator, and the number mentioned second is the denominator. The ratio 2:3 does not mean the same quantity as the ratio 3:2. Also, it is important to make sure than when discussing ratios that have units attached to them, the two quantities use the same units. For example, to compare of 8 feet to 4 yards, it would make sense to convert 4 yards to feet by multiplying by 3. Therefore, the ratio would be 8 feet to 12 feet, which can be expressed as the fraction $\frac{8}{12}$. Also, note that it is proper to refer to ratios in lowest terms. Therefore, the ratio of 8 feet to 4 yards is equivalent to the fraction $\frac{2}{3}$. Many real-world problems involve ratios. Often, problems with ratios involve proportions, as when two ratios are set equal to find the missing amount. However, some problems involve deciphering single ratios. For example, consider an amusement park that sold 345 tickets last Saturday. If 145 tickets were sold to adults and the rest of the tickets were sold to children, what would the ratio of the number of adult tickets to children's tickets be? A common mistake would be to say the ratio is 145:345. However, 345 is the total number of tickets sold. There were 345 – 145 = 200 tickets sold to children. The correct ratio of adult to children's tickets is 145:200. As a fraction, this expression is written as $\frac{145}{200}$, which can be reduced to $\frac{29}{40}$.

While a ratio compares two measurements using the same units, rates compare two measurements with different units. Examples of rates would be $200 for 8 hours of work, or 500 miles traveled per 20 gallons. Because the units are different, it is important to always include the units when discussing rates. Rates can be easily seen because if they are expressed in words, the two quantities are usually split up using one of the following words: *for, per, on, from, in*. Just as with ratios, it is important to write rates in lowest terms. A common rate that can be found in many real-life situations is cost per unit. This quantity describes how much one item or one unit costs. This rate allows the best buy to be determined, given a couple of different sizes of an item with different costs. For example, if 2 quarts of soup was sold for $3.50 and 3 quarts was sold for $4.60, to determine the best buy, the cost per quart should be found. $\frac{\$3.50}{2} = \$1.75$ per quart, and $\frac{\$4.60}{3} = \$1.53$ per quart. Therefore, the better deal would be the 3-quart option.

Rate of change problems involve calculating a quantity per some unit of measurement. Usually the unit of measurement is time. For example, meters per second is a common rate of change. To calculate this measurement, find the amount traveled in meters and divide by total time traveled. The calculation is an average of the speed over the entire time interval. Another common rate of change used in the real world is miles per hour. Consider the following problem that involves calculating an average rate of change in temperature. Last Saturday, the temperature at 1:00 a.m. was 34 degrees Fahrenheit, and at noon, the temperature had increased to 75 degrees Fahrenheit. What was the average rate of change over that time interval? The average rate of change is calculated by finding the change in temperature and dividing by the total hours elapsed. Therefore, the rate of change was equal to $\frac{75-34}{12-1} = \frac{41}{11}$ degrees per hour. This quantity rounded to two decimal places is equal to 3.72 degrees per hour.

A common rate of change that appears in algebra is the slope calculation. Given a linear equation in one variable, $y = mx + b$, the **slope**, m, is equal to $\frac{rise}{run}$ or $\frac{change\ in\ y}{change\ in\ x}$. In other words, slope is equivalent to the ratio of the vertical and horizontal changes between any two points on a line. The vertical change is known as the **rise**, and the horizontal change is known as the **run**. Given any two points on a line (x_1, y_1) and (x_2, y_2), slope can be calculated with the formula:

$$m = \frac{y_2 - y_1}{x_2 - x_1} = \frac{\Delta y}{\Delta x}$$

Common real-world applications of slope include determining how steep a staircase should be, calculating how steep a road is, and determining how to build a wheelchair ramp.

Many times, problems involving rates and ratios involve proportions. A proportion states that two ratios (or rates) are equal. The property of cross products can be used to determine if a proportion is true, meaning both ratios are equivalent. If $\frac{a}{b} = \frac{c}{d}$, then to clear the fractions, multiply both sides times the least common denominator, bd. This results in $ac = cd$, which is equal to the result of multiplying along both diagonals. For example, $\frac{4}{40} = \frac{1}{10}$ grants the cross product $4 \times 10 = 40 \times 1$. $40 = 40$ shows that this proportion is true. Cross products are used when proportions are involved in real-world problems. Consider the following: If 3 pounds of fertilizer will cover 75 square feet of grass, how many pounds are needed for 375 square feet? To solve this problem, a proportion can be set up using two ratios. Let x equal the unknown quantity, pounds needed for 375 feet. Then, the equation found by setting the two given ratios equal to one another is $\frac{3}{75} = \frac{x}{375}$. Cross multiplication gives $3 \times 375 = 75x$. Therefore, $1,125 = 75x$. Divide both sides by 75 to get $x = 15$. Therefore, 15 gallons of fertilizer is needed to cover 75 square feet of grass.

Another application of proportions involves similar triangles. If two triangles have the same measurement as two triangles in another triangle, the triangles are said to be **similar.** If two are the same, the third pair of angles are equal as well because the sum of all angles in a triangle is equal to 180 degrees. Each pair of equivalent angles are known as **corresponding angles. Corresponding sides** face the corresponding angles, and it is true that corresponding sides are in proportion.

For example, consider the following set of similar triangles:

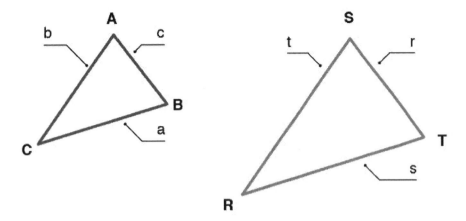

Angles A and R have the same measurement, angles C and T have the same measurement, and angles B and S have the same measurement. Therefore, the following proportion can be set up from the sides:

$$\frac{c}{t} = \frac{a}{r} = \frac{b}{s}$$

This proportion can be helpful in finding missing lengths in pairs of similar triangles. For example, if the following triangles are similar, a proportion can be used to find the missing side lengths, a and b.

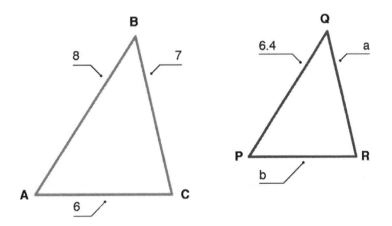

The proportions $\frac{8}{6.4} = \frac{6}{b}$ and $\frac{8}{6.4} = \frac{7}{a}$ can both be cross-multiplied and solved to obtain $a = 5.6$ and $b = 4.8$.

A real-life situation that uses similar triangles involves measuring shadows to find heights of unknown objects. Consider the following problem: A building casts a shadow that is 120 feet long, and at the same time, another building that is 80 feet high casts a shadow that is 60 feet long. How tall is the first building? Each building, together with the sun rays and shadows casted on the ground, forms a triangle. They are similar because each building forms a right angle with the ground, and the sun rays form equivalent angles. Therefore, these two pairs of angles are both equal. Because all angles in a triangle add up to 180 degrees, the third angles are equal as well. Both shadows form corresponding sides of the triangle, the buildings form corresponding sides, and the sun rays form corresponding sides. Therefore, the triangles are similar, and the following proportion can be used to find the missing building length:

$$\frac{120}{x} = \frac{60}{80}$$

Cross-multiply to obtain the cross products, $9600 = 60x$. Then, divide both sides by 60 to obtain $x = 160$. This solution means that the other building is 160 feet high.

Solving Unit Rate Problems

A **unit rate** is a rate with a denominator of one. It is a comparison of two values with different units where one value is equal to one. Examples of unit rates include 60 miles per hour and 200 words per minute. Problems involving unit rates may require some work to find the unit rate. For example, if Mary travels 360 miles in 5 hours, what is her speed, expressed as a unit rate? The rate can be expressed as the following fraction: $\frac{360 \; miles}{5 \; hours}$. The denominator can be changed to one by dividing by five. The numerator will also need to be divided by five to follow the rules of equality. This division turns the fraction into $\frac{72 \; miles}{1 \; hour}$, which can now be labeled as a unit rate because one unit has a value of one. Another type question involves the use of unit rates to solve problems. For example, if Trey needs to read 300 pages and his average speed is 75 pages per hour, will he be able to finish the reading in 5 hours? The unit rate is 75 pages per hour, so the total of 300 pages can be divided by 75 to find the time. After the division, the time it takes to read is four hours. The answer to the question is yes, Trey will finish the reading within 5 hours.

Recognizing Rational Exponents

A **rational number** is any number that can be written as a fraction of two integers. Examples of rational numbers include $\frac{1}{2}, \frac{5}{4}$, and 8. The number 8 is rational because it can be expressed as a fraction also, $\frac{8}{1} = 8$. **Rational exponents** represent one way to show how roots are used to express multiplication of any number by itself. For example, 3^2 has a base of 3 and rational exponent of 2, or $\frac{2}{1}$. It can be rewritten as the square root of 3 raised to the first power, or $\sqrt[2]{3^1}$. Any number with a rational exponent can be written this way. The **numerator,** or number on top of the fraction, becomes the root and the **denominator,** or bottom number of the fraction, becomes the whole number exponent. Another example is $4^{\frac{3}{2}}$. It can be rewritten as the square root of four to the third power, or $\sqrt[2]{4^3}$. This can be simplified by performing the operations 4 to the third power, $4^3 = 4 \times 4 \times 4 = 64$, and then taking the square root of 64, $\sqrt[2]{64}$, which yields an answer of 8. Another way of stating the answer would be 4 to power $\frac{3}{2}$ is eight, or 4 times itself $\frac{3}{2}$ times is eight.

Whole-Number Exponents

Numbers can also be written using exponents. The number 7,000 can be written as $7 \times 1,000$ because 7 is in the thousands place. It can also be written as 7×10^3 because $1,000 = 10^3$. Another number that can use this notation is 500. It can be written as 5×100, or 5×10^2, because $100 = 10^2$. The number 30 can be written as 3×10, or 3×10^1, because $10 = 10^1$. Notice that each one of the exponents of 10 is equal to the number of zeros in the number. Seven is in the thousands place, with three zeros, and the exponent on ten is 3. The five is in the hundreds place, with two zeros, and the exponent on the ten is 2. A question may give the number 40,000 and ask for it to be rewritten using exponents of ten. Because the number has a four in the ten-thousands place and four zeros, it can be written using an exponent of four: 4×10^4.

Understanding Vectors

A **vector** is something that has both magnitude and direction. A vector may sometimes be represented by a ray that has a length, for its magnitude, and a direction. As the magnitude of the vector increases, the length of the ray changes. The direction of the ray refers to the way that the magnitude is applied. The following vector shows the placement and parts of a vector:

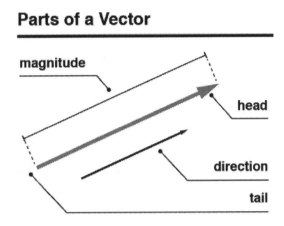

Examples of vectors include force and velocity. Force is a vector because applying force requires magnitude, which is the amount of force, and a direction, which is the way a force is applied. *Velocity* is a vector because it has a magnitude, or speed that an object travels, and also the direction that the object is going in. Vectors can be added together by placing the tail of the second at the head of the first. The resulting vector is found by starting at the first tail and ending at the second head. An example of this is show in the following picture:

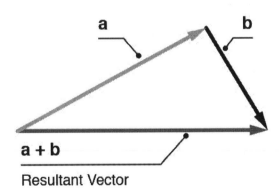

Subtraction can also be done with vectors by adding the inverse of the second vector. The inverse is found by reversing the direction of the vector. Then addition can take place just as described above, but using the inverse instead of the original vector. Scalar multiplication can also be done with vectors. This multiplication changes the magnitude of the vector by the *scalar*, or number. For example, if the length is described as 4, then scalar multiplication is used to multiply by 2, where the vector magnitude becomes 8. The direction of the vector is left unchanged because scalar does not include direction.

Vectors may also be described using coordinates on a plane, such as (5, 2). This vector would start at a point and move to the right 5 and up 2. The two coordinates describe the horizontal and vertical components of the vector. The starting point in relation to the coordinates is the tail, and the ending point is the head.

Creating Matrices

A **matrix** is an arrangement of numbers in rows and columns. Matrices are used to work with vectors and transform them. One example is a system of linear equations. Matrices can represent a system and be used to transform and solve the system. An important connection between scalars, vectors, and matrices is this: scalars are only numbers, vectors are numbers with magnitude and direction, and matrices are an array of numbers in rows and columns. The rows run from left to right and the columns run from top to bottom. When describing the dimensions of a matrix, the number of rows is stated first and the number of columns is stated second. The following matrix has two rows and three columns, referred to as a 2×3 matrix: $\begin{matrix} 3 & 5 & 7 \\ 4 & 2 & 8 \end{matrix}$. A number in a matrix can be found by describing its location. For example, the number in row two, column three is 8. In row one, column two, the number 5 is found.

Operations can be performed on matrices, just as they can on vectors. Scalar multiplication can be performed on matrices and it will change the magnitude, just as with a vector. A scalar multiplication problem using a 2×2 matrix looks like the following: $3 \times \begin{bmatrix} 4 & 5 \\ 8 & 3 \end{bmatrix}$. The scalar of 3 is multiplied by each number to form the resulting matrix: $\begin{bmatrix} 12 & 15 \\ 24 & 9 \end{bmatrix}$. Matrices can also be added and subtracted. For these operations to be performed, the matrices must be the same dimensions. Other operations that can be performed to manipulate matrices are multiplication, division, and transposition. *Transposing* a matrix means to switch the rows and columns. If the original matrix has two rows and three columns, then the transposed matrix has three rows and two columns.

Operations and Properties of Rational Numbers

Using Inverse Operations to Solve Problems

Inverse operations can be used to solve problems where there is a missing value. The area for a rectangle may be given, along with the length, but the width may be unknown. This situation can be modeled by the equation Area = Length × Width. The area is 40 square feet and the length is 10 feet. The equation becomes $40 = 10 \times w$. In order to find the w, we recognize that some number multiplied by 10 yields the number 40. The inverse operation to multiplication is division, so the 10 can be divided on both sides of the equation. This operation cancels out the 10 and yields an answer of 4 for the width. The following equation shows the work:

$$40 = 10 \, x \, w = \frac{40}{10} = \frac{10 \, x \, w}{10} = 4 = w$$

Other inverse operations can be used to solve problems as well. The following equation can be solved for b. $b + 4 = 9$. Because 4 is added to b, it can be subtracted on both sides of the equal sign to cancel out the four and solve for b, as follows:

$$b + 4 - 4 = 9 - 4 = b = 5$$

Whatever operation is used in the equation, the inverse operation can be used and applied to both sides of the equals sign to solve for an unknown value.

Interpreting Remainders in Division Problems

Understanding remainders begins with understanding the division problem. The problem $24 \div 7$ can be read as "twenty-four divided by seven." The problem is asking how many groups of 7 will fit into 24. Counting by seven, the multiples are 7, 14, 21, 28. Twenty-one, which is three groups of 7, is the closest to 24. The difference between 21 and 24 is 3, which is called the remainder. This is a remainder because it is the number that is left out after the three groups of seven are taken from 24. The answer to this division problem can be written as 3 with a remainder 3, or $3\frac{3}{7}$. The fraction $\frac{3}{7}$ can be used because it shows the part of the whole left when the division is complete. Another division problem may have the following numbers: $36 \div 5$. This problem is asking how many groups of 5 will fit evenly into 36. When counting by multiples of 5, the following list is generated: 5, 10, 15, 20, 25, 30, 35, 40. As seen in the list, there are six groups of five that make 35. To get to the total of 36, there needs to be one additional number. The answer to the division problem would be $36 \div 5 = 6R1$, or $6\frac{1}{5}$. The fractional part represents the number that cannot make up a whole group of five.

Performing Operations on Rational Numbers

Rational numbers are any numbers that can be written as a fraction of integers. Operations to be performed on rational numbers include adding, subtracting, multiplying, and dividing. Essentially, this refers to performing these operations on fractions. Adding and subtracting fractions must be completed by first finding the least common denominator. For example, the problem $\frac{3}{5} + \frac{6}{7}$ requires that the common multiple be found between 5 and 7. The smallest number that divides evenly by 5 and 7 is 35. For the denominators to become 35, they must be multiplied by 7 and 5 respectively. The fraction $\frac{3}{5}$ can be multiplied by 7 on the top and bottom to yield the fraction $\frac{21}{35}$. The fraction $\frac{6}{7}$ can be multiplied by 5 to yield the fraction $\frac{30}{35}$. Now that the fractions have the same denominator, the numerators can be added. The answer to the addition problem becomes

$$\frac{3}{5} + \frac{6}{7} = \frac{21}{35} + \frac{30}{35} = \frac{41}{35}$$

The same technique can be used for subtraction of rational numbers. The operations multiplication and division may seem easier to perform because finding common denominators is unnecessary. If the problems reads $\frac{1}{3} \times \frac{4}{5}$, then the numerators and denominators are multiplied by each other and the answer is found to be $\frac{4}{15}$. For division, the problem must be changed to multiplication before performing operations. The following words can be used to remember to leave, change, and flip before multiplying. If the problems reads $\frac{3}{7} \div \frac{3}{4}$, then the first fraction is *left* alone, the operation is *changed* to multiplication, and then the last fraction is *flipped*. The problem becomes

$$\frac{3}{7} \times \frac{4}{3} = \frac{12}{21}$$

Other rational numbers include negative numbers, where two may be added. When two negative numbers are added, the result is a negative number with an even greater magnitude. When a negative number is added to a positive number, the result depends on the value of each addend. For example, $-4 + 8 = 4$ because the positive number is larger than the negative number. For multiplying two negative numbers, the result is positive. For example, $-4 \times -3 = 12$, where the negatives cancel out and yield a positive answer.

Rational Numbers and Their Operations

Rational numbers can be whole or negative numbers, fractions, or repeating decimals because these numbers can all be written as a fraction. Whole numbers can be written as fractions, where 25 and 17 can be written as $\frac{25}{1}$ and $\frac{17}{1}$. One way of interpreting these fractions is to say that they are *ratios*, or comparisons of two quantities. The fractions given may represent 25 students to 1 classroom, or 17 desks to 1 computer lab. Repeating decimals can also be written as fractions of integers, such as 0.3333 and 0.6666667. These repeating decimals can be written as the fractions $\frac{1}{3}$ and $\frac{2}{3}$. Fractions can be described as having a part-to-whole relationship. The $\frac{1}{3}$ may represent 1 piece of pizza out of the whole cut into 3 pieces. The fraction $\frac{2}{3}$ may represent 2 pieces of the same whole pizza. One operation to perform on rational numbers, or fractions, can be addition. Adding the fractions $\frac{1}{3}$ and $\frac{2}{3}$ can be as simple as adding the numerators, 1 and 2. Because the denominator on both fractions is 3, that means the parts on the top have the same meaning, or are the same size piece of pizza. When adding these fractions, the result is $\frac{3}{3}$, or 1. Both of these numbers are rational and represent a whole, or in this problem, a whole pizza.

Other than fractions, rational numbers also include whole numbers and negative integers. When whole numbers are added, other than zero, the result is always greater than the addends. For example, the equation $4 + 18 = 22$ shows 4 increased by 18, with a result of 22. When subtracting rational numbers, sometimes the result is a negative number. For example, the equation $5 - 12 = -7$ shows that taking 12 away from 5 results in a negative answer because 5 is smaller than 12. The difference is $- 7$ because the starting number is smaller than the number taken away. For multiplication and division, similar results are found. Multiplying rational numbers may look like the following equation: $5 \times 7 = 35$, where both numbers are positive and whole, and the result is a larger number than the factors. The number 5 is counted 7 times, which results in a total of 35. Sometimes, the equation looks like $-4 \times 3 = -12$, so the result is negative because a positive number times a negative number gives a negative answer. The rule is that any time a negative number and a positive number are multiplied or divided, the result is negative.

Examples Where Multiplication Does Not Result in a Product Greater than Both Factors and Division Does Not Result in a Quotient Smaller than the Dividend
A common misconception of multiplication is that it always results in a value greater than the beginning number, or factors. This is not always the case. When working with fractions, multiplication may be used to take part of another number. For example, $\frac{1}{2} \times \frac{1}{4}$ can be read as "one-half times one-fourth," or taking one-half of one-fourth. The latter translation makes it easier to understand the concept. Taking half of one-fourth will result in a smaller number that one-fourth. It will result in one-eighth. The same happens with multiplying two-thirds times three-fifths, or $\frac{2}{3} \times \frac{3}{5}$. The concept of taking two-thirds, which is a part, of three-fifths, means that there will be an even smaller part as the result. Multiplication of these two fractions yields the answer $\frac{6}{15}$, or $\frac{2}{5}$.

In the same way, another misconception is that division always has results smaller than the beginning number or dividend. When working with whole numbers, division asks how many times a whole goes into another whole. This result will always be smaller than the dividend, where $6 \div 2 = 3$ and $20 \div 5 = 4$. When working with fractions, the number of times a part goes into another part depends on the value of each fraction. For example, three-fourths divided by one-fourth, or $\frac{3}{4} \div \frac{1}{4}$, asks to find how many times $\frac{1}{4}$ will go into $\frac{3}{4}$. Because these have the same denominator, the numerators can be compared as is, without needing to convert the fractions. The result is easily found to be 3 because one goes into three 3 times.

Composing and Decomposing Fractions

Fractions are ratios of whole numbers and their negatives. Fractions represent parts of wholes, whether pies, or money, or work. The number on top, or numerator, represents the part, and the bottom number, or denominator, represents the whole. The number $\frac{1}{2}$ represents half of a whole. Other ways to represent one-half are $\frac{2}{4}, \frac{3}{6}$, and $\frac{5}{10}$. These are fractions not written in simplest form, but the numerators are all halves of the denominators. The fraction $\frac{1}{4}$ represents 1 part to a whole of 4 parts. This can be seen by the quarter's value in relation to the dollar. One quarter is $\frac{1}{4}$ of a dollar. In the same way, 2 quarters make up $\frac{1}{2}$ of a dollar, so 2 fractions of $\frac{1}{4}$ make up a fraction of $\frac{1}{2}$. Three quarters make up three-fourths of a dollar. The three fractions of $\frac{1}{4} + \frac{1}{4} + \frac{1}{4}$ are equal to $\frac{3}{4}$ of a whole. This illustration can be seen using the bars below divided into one whole, then two halves, then three sections of one-third, then four sections of one-fourth. Based on the size of the fraction, different numbers of each fraction are needed to make up a whole.

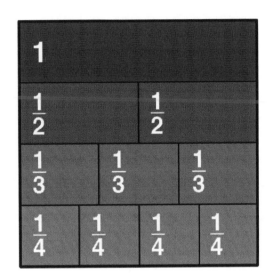

The Value of a Unit Fraction

A **unit fraction** is a fraction where the numerator has a value of one. The fractions one-half, one-third, one-seventh, and one-tenth are all examples of unit fractions. Examples that are not unit fractions include three-fourths, four-fifths, and seven-twelfths. The value of unit fractions changes as the denominator changes, because the numerator is always one. The unit fraction one-half requires two parts to make a whole. The unit fraction one-third requires three parts to make a whole. In the same way, if the unit fraction changes to one-thirteenth, then the number of parts required to make a whole becomes thirteen. An illustration of this is seen in the figure below. As the denominator increases, the size of the parts for each fraction decreases. As the bar goes from one-fourth to one-fifth, the size of the bars decreases, but the size of the denominator increases to five.

This pattern continues down the diagram as the bars, or value of the fraction, get smaller, the denominator gets larger:

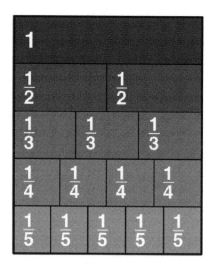

Using the Same Whole When Comparing Fractions

Comparing fractions requires the use of a common denominator. The necessity of this can be seen by the two pies below. The first pie has a shaded value of $2/10$ because two pieces are shaded out of the whole of ten pieces. The second pie has a shaded value of $2/7$ because two pieces are shaded out of a whole of seven pieces. These two fractions, two-tenths and two-sevenths, have the same numerator and so a misconception may be that they are equal. By looking at the shaded region in each pie, it is apparent that the fractions are not equal. The numerators are the same, but the denominators are not. Two parts of a whole are not equivalent unless the whole is broken into the same number of parts. To compare the shaded regions, the denominators seven and ten must be made equal. The lowest number that the two denominators will both divide evenly into is 70, which is the lowest common denominator. Then the numerators must be converted by multiplying by the opposite denominator. These operations result in the two fractions $14/70$ and $20/70$. Now that these two have the same denominator, the conclusion can be made that $2/7$ represents a larger portion of the pie, as seen in the figure below:

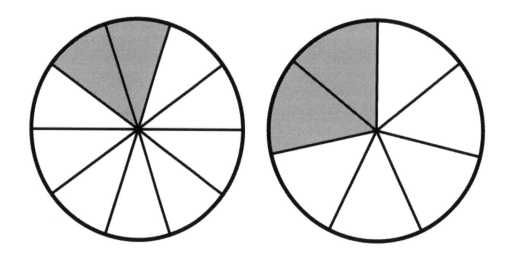

The Order of Operations

As mentioned, the **order of operations** refers to the order in which problems are to be solved, from parenthesis or grouping, to addition and subtraction. A common way of remembering the order of operations is PEMDAS, or "Please Excuse My Dear Aunt Sally." The letters stand for parenthesis, or grouping, exponents, multiply/divide, and add/subtract. The first step is to complete any operations inside the grouping symbols, or parenthesis. The next step is to simplify all exponents. After exponents, the operations of multiplication and division are performed in the order they appear from left to right. The last operations are addition and subtraction, also performed from left to right. The following problem requires the use of order of operations to be solved: $2(3^3 + 5) - 8$. The first step is to perform the operations inside the grouping symbols, or parenthesis. Inside the parenthesis, the exponent would be performed first, then the addition of $(3^3 + 5)$ which is $(27 + 5)$ or (32). These operations lead to the next step of $2(32) - 8$, where the multiplication can be performed between 2 and 8. This step leads to the problem $64 - 8$, where the answer is 56. The order of operations is important because if solved in a different order, the resulting number would not be 56. A common of when the order of operations can be used is when a store is having a sale and customers may use coupons. Other places may be at a restaurant, for the check, or the gas station when using a card to pay.

Representing Rational Numbers and Their Operations in Different Ways

Rational numbers can be written as fractions, but also as a percent or decimal. For example, three-fourths is a fraction written as three divided by four or $\frac{3}{4}$. It represents three parts out of a whole of four parts. By dividing three by four, the decimal of 0.75 is found. This decimal represents the same part to whole relationship as three-fourths. Seventy-five is in the hundredths place, so it can be read as 75 out of 100, the same ratio as 3 to 4. The decimal 0.75 is the same as 75 out of 100, or 75%. The rational number three-fourths represents the same portion as the decimal 0.75 and the percentage 75%. Because there are different ways to represent rational numbers, the operations used to manipulate rational numbers can look different also. For the operation of multiplication, the problem can use a dot, an "x," or simply writing two variables side by side to indicate the need to find the product. Division can be represented by the line in a fraction or a division symbol. When adding or subtracting, the form of rational numbers is important and can be changed. Sometimes it is simpler to work with fractions, while sometimes decimals are easier to manipulate, depending on the operation. When comparing portion size, it may be easier to see each number as a percent, so it is important to understand the different ways rational numbers can be represented.

Representing Rational Numbers on a Number Line

A **number line** is a tool used to compare numbers by showing where they fall in relation to one another. Labeling a number line with integers is simple because they have no fractional component and the values are easier to understand. The number line may start at -3 and go up to -2, then -1, then 0, and 1, 2, 3. This order shows that number 2 is larger than -1 because it falls further to the right on the number line. When positioning rational numbers, the process may take more time because it requires that they all be in the same form. If they are percentages, fractions, and decimals, then conversions will have to be made to put them in the same form. For example, if the numbers $\frac{5}{4}$, 45%, and 2.38 need to be put in order on a number line, the numbers must first be transformed into one single form. Decimal form is an easy common ground because fractions can be changed by simply dividing and percentages can be changed by moving the decimal point. After conversions are made, the list becomes 1.25, 0.45, and 2.38 respectively. Now the list is easier to arrange.

The number line with the list in order is shown in the top half of the graphic below in the order 0.45, 1.25, and 2.38:

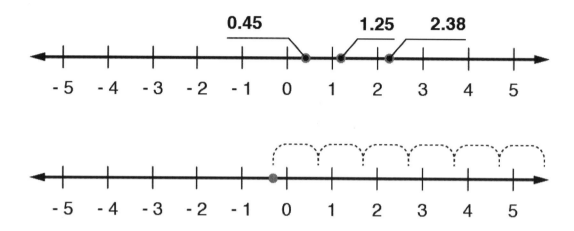

The sums and differences of rational numbers can be found using a number line after the rational numbers are put into the same form. This method is especially helpful when understanding the addition and subtraction of negative numbers. For example, the rational number six can be added to negative one-half using the number line. The following expression represents the problem: $-\frac{1}{2} + 6$. First, the original number $-\frac{1}{2}$ can be labeled as in the lower half of the graphic above, by the dot. Then 6 can be added by counting by whole numbers on the number line. The arcs on the graph represent the addition. The final answer is positive 5 ½.

Illustrating Multiplication and Division Problems Using Equations, Rectangular Arrays, and Area Models

Multiplication and division can be represented by equations. These equations show the numbers involved and the operation. For example, "eight multiplied by six equals forty-eight" is seen in the following equation: $8 \times 6 = 48$. This operation can be modeled by rectangular arrays where one factor, 8, is the number of rows, and the other factor, 6, is the number of columns, as follows:

Array of 8 x 6 = 48

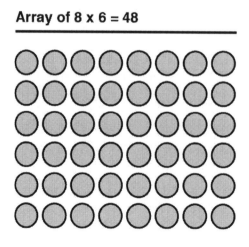

Rectangular arrays show what happens with the concept of multiplication. As one row of dots is drawn, that represents the first factor in the problem. Then the second factor is used to add the number of columns. The final model includes six rows of eight columns which results in forty-eight dots. These rectangular arrays show how multiplication of whole numbers will result in a number larger than the factors.

Division can also be represented by equations and area models. A division problem such as "twenty-four divided by three equals eight" can be written as the following equation: $24 \div 8 = 3$. The object below shows an area model to represent the equation. As seen in the model, the whole box represents 24 and the 3 sections represent the division by 3. In more detail, there could be 24 dots written in the whole box and each box could have 8 dots in it. Division shows how numbers can be divided into groups. For the example problem, it is asking how many numbers will be in each of the 3 groups that together make 24. The answer is 8 in each group.

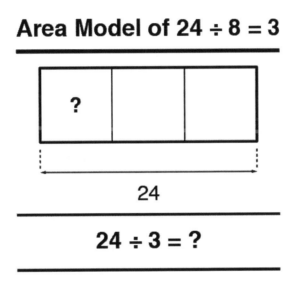

Area Model of 24 ÷ 8 = 3

$$24 \div 3 = ?$$

Practice Questions

1. What is $\frac{12}{60}$ converted to a percentage?
    a. 0.20
    b. 20%
    c. 25%
    d. 12%
    e. 1.2%

2. Which of the following is the correct decimal form of the fraction $\frac{14}{33}$ rounded to the nearest hundredth place?
    a. 0.420
    b. 0.14
    c. 0.424
    d. 0.140
    e. 0.42

3. Which of the following represents the correct sum of $\frac{14}{15}$ and $\frac{2}{5}$?
    a. $\frac{20}{15}$
    b. $\frac{4}{3}$
    c. $\frac{16}{20}$
    d. $\frac{4}{5}$
    e. $\frac{16}{15}$

4. What is the product of $\frac{5}{14}$ and $\frac{7}{20}$?
    a. $\frac{1}{8}$
    b. $\frac{35}{280}$
    c. $\frac{12}{34}$
    d. $\frac{1}{2}$
    e. $\frac{7}{140}$

5. What is the result of dividing 24 by $\frac{8}{5}$?

   a. $\frac{5}{3}$

   b. $\frac{3}{5}$

   c. $\frac{120}{8}$

   d. 15

   e. $\frac{24}{5}$

6. Subtract $\frac{5}{14}$ from $\frac{5}{24}$. Which of the following is the **correct** result?

   a. $\frac{25}{168}$

   b. 0

   c. $-\frac{25}{168}$

   d. $\frac{1}{10}$

   e. $-\frac{1}{10}$

7. Which of the following is a **correct** mathematical statement?

   a. $\frac{1}{3} < -\frac{4}{3}$

   b. $-\frac{1}{3} > \frac{4}{3}$

   c. $\frac{1}{3} > \frac{4}{3}$

   d. $-\frac{1}{3} \geq \frac{4}{3}$

   e. $\frac{1}{3} > -\frac{4}{3}$

8. Which of the following is **incorrect**?

   a. $-\frac{1}{5} < \frac{4}{5}$

   b. $\frac{4}{5} > -\frac{1}{5}$

   c. $-\frac{1}{5} > \frac{4}{5}$

   d. $\frac{1}{5} > -\frac{4}{5}$

   e. $\frac{4}{5} > \frac{1}{5}$

9. In a neighborhood, 15 out of 80 of the households have children under the age of 18. What percentage of the households have children?

    a. 0.1875%

    b. 18.75%

    c. 1.875%

    d. 15%

    e. 1.50%

10. Gina took an algebra test last Friday. There were 35 questions, and she answered 60% of them correctly. How many correct answers did she have?

    a. 35

    b. 20

    c. 21

    d. 25

    e. 18

11. Paul took a written driving test, and he got 12 of the questions correct. If he answered 75% of the total questions correctly, how many problems were on the test?

    a. 25

    b. 15

    c. 20

    d. 18

    e. 16

12. If a car is purchased for $15,395 with a 7.25% sales tax, what is the total price?

    a. $15,395.07

    b. $16,511.14

    c. $16,411.13

    d. $15,402

    e. $16,113.10

13. A car manufacturer usually makes 15,412 SUVs, 25,815 station wagons, 50,412 sedans, 8,123 trucks, and 18,312 hybrids a month. About how many cars are manufactured each month?

    a. 120,000

    b. 200,000

    c. 300,000

    d. 12,000

    e. 20,000

14. Each year, a family goes to the grocery store every week and spends $105. About how much does the family spend annually on groceries?

    a. $10,000

    b. $50,000

    c. $500

    d. $5,000

    e. $1,200

15. Bindee is having a barbeque on Sunday and needs 12 packets of ketchup for every 5 guests. If 60 guests are coming, how many packets of ketchup should she buy?

    a. 100

    b. 12

    c. 144

    d. 60

    e. 300

16. A grocery store sold 48 bags of apples in one day, and 9 of the bags contained Granny Smith apples. The rest contained Red Delicious apples. What is the ratio of bags of Granny Smith to bags of Red Delicious that were sold?

    a. 48:9

    b. 39:9

    c. 9:48

    d. 9:39

    e. 39:48

17. If Oscar's bank account totaled $4,000 in March and $4,900 in June, what was the rate of change in his bank account total over those three months?

    a. $900 a month

    b. $300 a month

    c. $4,900 a month

    d. $100 a month

    e. $4,000 a month

18. How is a transposition of a matrix performed?

    a. Multiply each number by negative 1

    b. Switch the rows and columns

    c. Reverse the order of each row

    d. Find the inverse of each number

    e. Divide the first number in each row by the last number in the last column

19. What is the label given to a problem that multiplies a matrix by a constant?

    a. Vector multiplication

    b. Scalar multiplication

    c. Inverse of a matrix

    d. Transposition of a matrix

    e. Product of a matrix

20. What are the two values that always describe a vector?

    a. Magnitude and direction

    b. Magnitude and length

    c. Length and position

    d. Direction and position

    e. Magnitude and position

21. What is the resultant matrix when addition is performed?

$$\begin{bmatrix} 1 & -5 \\ 4 & 2 \end{bmatrix} + \begin{bmatrix} 3 & 2 \\ -2 & 1 \end{bmatrix} = \begin{bmatrix} & \\ & \end{bmatrix}$$

a. $\begin{bmatrix} 3 & -2 \\ 5 & 0 \end{bmatrix}$

b. $\begin{bmatrix} 4 & 3 \\ 2 & 3 \end{bmatrix}$

c. $\begin{bmatrix} 4 & -3 \\ 2 & 3 \end{bmatrix}$

d. $\begin{bmatrix} 3 & -3 \\ -3 & 3 \end{bmatrix}$

e. $\begin{bmatrix} 3 & 3 \\ 2 & 3 \end{bmatrix}$

22. Which set of matrices represents the following system of equations?

$$\begin{cases} x - 2y + 3z = 7 \\ 2x + y + z = 4 \\ -3x + 2y - 2z = -10 \end{cases}$$

a. $\begin{bmatrix} 1 & -2 & 3 \\ 2 & 3 & 1 \\ -3 & 1 & 2 \end{bmatrix} \begin{bmatrix} x \\ y \\ z \end{bmatrix} = \begin{bmatrix} 7 \\ -4 \\ 10 \end{bmatrix}$

b. $\begin{bmatrix} 1 & 2 & -3 \\ -2 & 1 & 2 \\ 3 & 1 & -2 \end{bmatrix} \begin{bmatrix} x \\ y \\ z \end{bmatrix} = \begin{bmatrix} 7 \\ 4 \\ -10 \end{bmatrix}$

c. $\begin{bmatrix} 1 & -2 & 4 \\ -3 & 1 & 7 \\ 2 & 2 & -10 \end{bmatrix} \begin{bmatrix} 3 \\ 1 \\ -2 \end{bmatrix} = \begin{bmatrix} x \\ y \\ z \end{bmatrix}$

d. $\begin{bmatrix} 1 & -2 & 3 \\ 2 & 1 & 1 \\ -3 & 2 & -2 \end{bmatrix} \begin{bmatrix} x \\ y \\ z \end{bmatrix} = \begin{bmatrix} 7 \\ 4 \\ -10 \end{bmatrix}$

e. $\begin{bmatrix} -1 & 2 & -3 \\ 2 & 1 & 1 \\ -3 & 2 & 2 \end{bmatrix} \begin{bmatrix} x \\ y \\ z \end{bmatrix} = \begin{bmatrix} 5 \\ 4 \\ -8 \end{bmatrix}$

23. Which label describes the relationship between these vectors?

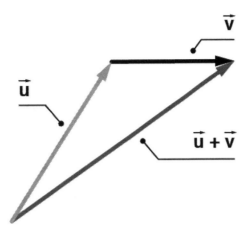

    a. Vector magnitude
    b. Vector extraction
    c. Scalar vectors
    d. Vector addition
    e. Scalar magnitude

24. What is the name of the indicated part of Vector B?

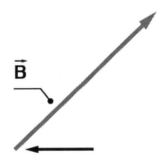

    a. Head
    b. Tail
    c. Magnitude
    d. Direction
    e. Point

25. When two vectors are added, what is the answer called?
    a. Ending vector
    b. Total vector
    c. Summation vector
    d. Resolved vector
    e. Resultant vector

Answer Explanations

1. B: The fraction $\frac{12}{60}$ can be reduced to $\frac{1}{5}$, in lowest terms. First, it must be converted to a decimal. Dividing 1 by 5 results in 0.2. Then, to convert to a percentage, move the decimal point two units to the right and add the percentage symbol. The result is 20%.

2. E: If a calculator is used, divide 33 into 14 and keep two decimal places. If a calculator is not used, multiply both the numerator and denominator times 3. This results in the fraction $\frac{42}{99}$, and hence an approximate decimal of 0.42.

3. B: Common denominators must be used. The LCD is 15, and $\frac{2}{5} = \frac{6}{15}$. Therefore, $\frac{14}{15} + \frac{6}{15} = \frac{20}{15}$, and in lowest terms, the answer is $\frac{4}{3}$. A common factor of 5 was divided out of both the numerator and denominator.

4. A: A product is found by multiplication. Multiplying two fractions together is easier when common factors are cancelled first to avoid working with larger numbers.

$$\frac{5}{14} \times \frac{7}{20} = \frac{5}{2 \times 7} \times \frac{7}{5 \times 4}$$

$$\frac{1}{2} \times \frac{1}{4} = \frac{1}{8}$$

5. D: Division is completed by multiplying by the reciprocal. Therefore:

$$24 \div \frac{8}{5} = \frac{24}{1} \times \frac{5}{8}$$

$$\frac{3 \times 8}{1} \times \frac{5}{8} = \frac{15}{1} = 15$$

6. C: Common denominators must be used. The LCD is 168, so each fraction must be converted to have 168 as the denominator.

$$\frac{5}{24} - \frac{5}{14} = \frac{5}{24} \times \frac{7}{7} - \frac{5}{14} \times \frac{12}{12}$$

$$\frac{35}{168} - \frac{60}{168} = -\frac{25}{168}$$

7. E: The correct mathematical statement is the one in which the number to the left on the number line is less than the number to the right on the number line. It is written in answer E that $\frac{1}{3} > -\frac{4}{3}$, which is the same as $-\frac{4}{3} < \frac{1}{3}$, a correct statement.

8. C: $-\frac{1}{5} > \frac{4}{5}$ is incorrect. The expression on the left is negative, which means that it is smaller than the expression on the right. As it is written, the inequality states that the expression on the left is greater than the expression on the right, which is not true.

9. B: First, the information is translated into the ratio $\frac{15}{80}$. To find the percentage, translate this fraction into a decimal by dividing 15 by 80. The corresponding decimal is 0.1875. Move the decimal point two units to the right to obtain the percentage 18.75%.

10. C: Gina answered 60% of 35 questions correctly; 60% can be expressed as the decimal 0.60. Therefore, she answered $0.60 \times 35 = 21$ questions correctly.

11. E: The unknown quantity is the number of total questions on the test. Let x be equal to this unknown quantity. Therefore, $0.75x = 12$. Divide both sides by 0.75 to obtain $x = 16$.

12. B: If sales tax is 7.25%, the price of the car must be multiplied times 1.0725 to account for the additional sales tax. Therefore, $15,395 \times 1.0725 = 16,511.1375$. This amount is rounded to the nearest cent, which is $16,511.14.

13. A: Rounding can be used to find the best approximation. All of the values can be rounded to the nearest thousand. 15,412 SUVs can be rounded to 15,000. 25,815 station wagons can be rounded to 26,000. 50,412 sedans can be rounded to 50,000. 8,123 trucks can be rounded to 8,000. Finally, 18,312 hybrids can be rounded to 18,000. The sum of the rounded values is 117,000, which is closest to 120,000.

14. D: There are 52 weeks in a year, and if the family spends $105 each week, that amount is close to $100. A good approximation is $100 a week for 50 weeks, which is found through the product $50 \times 100 = \$5,000$.

15. C: This problem involves ratios and percentages. If 12 packets are needed for every 5 people, this statement is equivalent to the ratio $\frac{12}{5}$. The unknown amount x is the number of ketchup packets needed for 60 people. The proportion $\frac{12}{5} = \frac{x}{60}$ must be solved. Cross-multiply to obtain $12 \times 60 = 5x$. Therefore, $720 = 5x$. Divide each side by 5 to obtain $x = 144 \ packets$.

16. D: There were 48 total bags of apples sold. If 9 bags were Granny Smith and the rest were Red Delicious, then $48 - 9 = 39$ bags were Red Delicious. Therefore, the ratio of Granny Smith to Red Delicious is 9:39.

17. B: The average rate of change is found by calculating the difference in dollars over the elapsed time. Therefore, the rate of change is equal to $4,900-$4,000÷3 months, which is equal to $900÷3 or $300 a month.

18. B: The correct choice is *B* because the definition of transposing a matrix says that the rows and columns should be switched. For example, the matrix $\begin{bmatrix} 3 & 4 \\ 2 & 5 \\ 1 & 6 \end{bmatrix}$ can be transposed into $\begin{bmatrix} 3 & 2 & 1 \\ 4 & 5 & 6 \end{bmatrix}$. Notice that the first row, 3 and 4, becomes the first column. The second row, 2 and 5, becomes the second column. This is an example of transposing a matrix.

19. B: The correct answer is Choice *B* because multiplying a matrix by a constant is called scalar multiplication. A scalar is a constant number, which means the only thing it changes about a matrix is its magnitude. For a given matrix, $\begin{bmatrix} 3 & 4 \\ 6 & 5 \end{bmatrix}$, scalar multiplication can be applied by multiplying by 2, which yields the matrix $\begin{bmatrix} 6 & 8 \\ 12 & 10 \end{bmatrix}$. Notice that the dimensions of the matrix did not change, just the magnitude of the numbers.

20. A: The vector is described as having both magnitude and direction. The magnitude is the size of the vector and the direction is the path along with which the force is being applied. The second answer choice has magnitude and length, which are essentially the same. The third, fourth, and fifth answer choices include length and position, but position is also not part of the description of a vector.

21. C: The correct answer is Choice *C* because matrices are added by adding the corresponding numbers in the two given matrices. For matrices to be added, they must have the same number of rows and columns. These two matrices have two rows and two columns. The first row, first column numbers are 1 and 3, and adding them together yields a total of 4. The second row, first column has numbers 4 and –2, which yields a sum of 2. The first row, second column has –5 and 2, which sum to –3. The second row, second column have 2 and 1, which yield a sum of 3. The resulting matrix after addition is: $\begin{bmatrix} 4 & -3 \\ 2 & 3 \end{bmatrix}$.

$$\begin{bmatrix} 1 & -5 \\ 4 & 2 \end{bmatrix} + \begin{bmatrix} 3 & 2 \\ -2 & 1 \end{bmatrix} = \begin{bmatrix} 4 & -3 \\ 2 & 3 \end{bmatrix}$$

22. D: The correct matrix to describe the given system of equations is the last matrix (Choice *D*) because it has values that correspond to the coefficients in the right order. The top row corresponds to the coefficient in the first equation, the second row corresponds to the coefficients in the second equation, and the third row corresponds to the coefficients in the third equation. The second matrix (Choice *B*) is filled with the three variables in the system. One thing to also look for is the sign on the numbers, to make sure the signs are correct from the equation to the matrix.

23. D: The picture shows how vectors are added. According to the definition, the vectors must be placed head to tail, then the resulting vector is drawn from the beginning head to the ending tail. Notice that the vector runs from the bottom left to the top right. Adding these vectors in the wrong direction would yield a vector starting at the top right, at the end of vector *v*, and ending at the bottom left.

24. B: The answer is Choice *B* because the part of the vector that the arrow points to is the tail. The end of a vector with the arrow is called the head. The opposite end, without the arrow, is called the tail.

25. E: When two vectors are added, the resulting vector is called the resultant. By definition, the vector found from addition of two or more vectors is the resultant. The other choices of ending vector and total vector could refer to one or more of the starting vectors, not the resultant.

Word Knowledge

Word Knowledge

The Word Knowledge section of the AFOQT is designed to assess the test taker's vocabulary knowledge and skill in determining the answer choice with a meaning that most nearly matches that of the word presented in the question. In this way, Word Knowledge questions task test takers with applying their vocabulary skills to select the best synonyms.

Question Format

The questions in the Word Knowledge section are constructed very simply: the prompt is a single word, which is then followed by the five answer choices, each of which is also a single word. Test takers must consider the definition or meaning of the word provided in the prompt, and then select the answer choice that most nearly means the same thing. Consider the following example of a Word Knowledge question:

SERENE
    a. Serious
    b. Calm
    c. Tired
    d. Nervous
    e. Jealous

Test takers must consider the meaning of the given word (*serene*), read the five potential definitions or synonyms, and pick the word that most closely means the same thing as the word in the prompt. In this case, Choice *B* is the best answer because *serene* means calm. The format of this example question is exactly like all of the 25 questions that will appear in the Word Knowledge section of the AFOQT.

Denotation and Connotation

Denotation refers to a word's explicit definition, like that found in the dictionary. Denotation is often set in comparison to connotation. **Connotation** is the emotional, cultural, social, or personal implication associated with a word. Denotation is more of an objective definition, whereas connotation can be more subjective, although many connotative meanings of words are similar for certain cultures. The denotative meanings of words are usually based on facts, and the connotative meanings of words are usually based on emotion.

Here are some examples of words and their denotative and connotative meanings in Western culture:

Word	Denotative Meaning	Connotative Meaning
Home	A permanent place where one lives, usually as a member of a family.	A place of warmth; a place of familiarity; comforting; a place of safety and security. "Home" usually has a positive connotation.
Snake	A long reptile with no limbs and strong jaws that moves along the ground; some snakes have a poisonous bite.	An evil omen; a slithery creature (human or nonhuman) that is deceitful or unwelcome. "Snake" usually has a negative connotation.
Winter	A season of the year that is the coldest, usually from December to February in the northern hemisphere and from June to August in the southern hemisphere.	Circle of life, especially that of death and dying; cold or icy; dark and gloomy; hibernation, sleep, or rest. "Winter" can have a negative connotation, although many who have access to heat may enjoy the snowy season from their homes.

Analyzing Word Parts

By learning some of the etymologies of words and their parts, readers can break new words down into components and analyze their combined meanings. For example, the root word *soph* is Greek for wise or knowledge. Knowing this informs the meanings of English words including *sophomore, sophisticated,* and *philosophy.* Those who also know that *phil* is Greek for love will realize that *philosophy* means the love of knowledge. They can then extend this knowledge of *phil* to understand *philanthropist* (one who loves people), *bibliophile* (book lover), *philharmonic* (loving harmony), *hydrophilic* (water-loving), and so on. In addition, *phob-* derives from the Greek *phobos,* meaning fear. This informs all words ending with it as meaning fear of various things: *acrophobia* (fear of heights), *arachnophobia* (fear of spiders), *claustrophobia* (fear of enclosed spaces), *ergophobia* (fear of work), and *hydrophobia* (fear of water), among others.

Some words that originate from other languages, like ancient Greek, are found in large numbers and varieties of English words. An advantage of the shared ancestry of these words is that once readers recognize the meanings of some Greek words or word roots, they can determine or at least get an idea of what many different English words mean. As an example, the Greek word *métron* means to measure, a measure, or something used to measure; the English word meter derives from it. Knowing this informs many other English words, including *altimeter, barometer, diameter, hexameter, isometric,* and *metric.* While readers must know the meanings of the other parts of these words to decipher their meaning fully, they already have an idea that they are all related in some way to measures or measuring.

While all English words ultimately derive from a proto-language known as Indo-European, many of them historically came into the developing English vocabulary later, from sources like the ancient Greeks' language, the Latin used throughout Europe and much of the Middle East during the reign of the Roman Empire, and the Anglo-Saxon languages used by England's early tribes. In addition to classic revivals and native foundations, by the Renaissance era, other influences included French, German, Italian, and Spanish. Today we can often discern English word meanings by knowing common roots and affixes, particularly from Greek and Latin.

The following is a list of common prefixes and their meanings:

Prefix	Definition	Examples
a-	without	atheist, agnostic
ad-	to, toward	advance
ante-	before	antecedent, antedate
anti-	opposing	antipathy, antidote
auto-	self	autonomy, autobiography
bene-	well, good	benefit, benefactor
bi-	two	bisect, biennial
bio-	life	biology, biosphere
chron-	time	chronometer, synchronize
circum-	around	circumspect, circumference
com-	with, together	commotion, complicate
contra-	against, opposing	contradict, contravene
cred-	belief, trust	credible, credit
de-	from	depart
dem-	people	demographics, democracy
dis-	away, off, down, not	dissent, disappear
equi-	equal, equally	equivalent
ex-	former, out of	extract
for-	away, off, from	forget, forswear
fore-	before, previous	foretell, forefathers
homo-	same, equal	homogenized
hyper-	excessive, over	hypercritical, hypertension
in-	in, into	intrude, invade
inter-	among, between	intercede, interrupt
mal-	bad, poorly, not	malfunction
micr-	small	microbe, microscope
mis-	bad, poorly, not	misspell, misfire
mono-	one, single	monogamy, monologue
mor-	die, death	mortality, mortuary
neo-	new	neolithic, neoconservative
non-	not	nonentity, nonsense
omni-	all, everywhere	omniscient
over-	above	overbearing
pan-	all, entire	panorama, pandemonium
para-	beside, beyond	parallel, paradox
phil-	love, affection	philosophy, philanthropic
poly-	many	polymorphous, polygamous
pre-	before, previous	prevent, preclude

prim-	first, early	primitive, primary
pro-	forward, in place of	propel, pronoun
re-	back, backward, again	revoke, recur
sub-	under, beneath	subjugate, substitute
super-	above, extra	supersede, supernumerary
trans-	across, beyond, over	transact, transport
ultra-	beyond, excessively	ultramodern, ultrasonic, ultraviolet
un-	not, reverse of	unhappy, unlock
vis-	to see	visage, visible

The following is a list of common suffixes and their meanings:

Suffix	Definition	Examples
-able	likely, able to	capable, tolerable
-ance	act, condition	acceptance, vigilance
-ard	one that does excessively	drunkard, wizard
-ation	action, state	occupation, starvation
-cy	state, condition	accuracy, captaincy
-er	one who does	teacher
-esce	become, grow, continue	convalesce, acquiesce
-esque	in the style of, like	picturesque, grotesque
-ess	feminine	waitress, lioness
-ful	full of, marked by	thankful, zestful
-ible	able, fit	edible, possible, divisible
-ion	action, result, state	union, fusion
-ish	suggesting, like	churlish, childish
-ism	act, manner, doctrine	barbarism, socialism
-ist	doer, believer	monopolist, socialist
-ition	action, result, state,	sedition, expedition
-ity	quality, condition	acidity, civility
-ize	cause to be, treat with	sterilize, mechanize, criticize
-less	lacking, without	hopeless, countless
-like	like, similar	childlike, dreamlike
-ly	like, of the nature of	friendly, positively
-ment	means, result, action	refreshment, disappointment
-ness	quality, state	greatness, tallness
-or	doer, office, action	juror, elevator, honor
-ous	marked by, given to	religious, riotous
-some	apt to, showing	tiresome, lonesome
-th	act, state, quality	warmth, width
-ty	quality, state	enmity, activity

The following is a list of root words and their meanings:

Root	Definition	Examples
ambi	both	ambidextrous, ambiguous
anthropo	man; humanity	anthropomorphism, anthropology
auto	self	automobile, autonomous
bene	good	benevolent, benefactor
bio	life	biology, biography
chron	time	chronology
circum	around	circumvent, circumference
dyna	power	dynasty, dynamite
fort	strength	fortuitous, fortress
graph	writing	graphic
hetero	different	heterogeneous
homo	same	homonym, homogenous
hypo	below, beneath	hypothermia
morph	shape; form	morphology
mort	death	mortal, mortician
multi	many	multimedia, multiplication
nym	name	antonym, synonym
phobia	fear	claustrophobia
port	carry	transport
pseudo	false	pseudoscience, pseudonym
scope	viewing instrument	telescope, microscope
techno	art; science; skill	technology, techno
therm	heat	thermometer, thermal
trans	across	transatlantic, transmit
under	too little	underestimate

Practice Questions

1. WARY
   a. Religious
   b. Adventurous
   c. Tired
   d. Negligent
   e. Cautious

2. PROXIMITY
   a. Estimate
   b. Delicate
   c. Precarious
   d. Splendor
   e. Closeness

3. PARSIMONIOUS
   a. Lavish
   b. Harmonious
   c. Miserly
   d. Careless
   e. Generous

4. PROPRIETY
   a. Ownership
   b. Appropriateness
   c. Patented
   d. Abstinence
   e. Sobriety

5. BOON
   a. Cacophony
   b. Hopeful
   c. Benefit
   d. Squall
   e. Omen

6. STRIFE
   a. Plague
   b. Industrial
   c. Picketing
   d. Eliminate
   e. Conflict

7. QUALM
    a. Calm
    b. Uneasiness
    c. Assertion
    d. Pacify
    e. Victory

8. VALOR
    a. Rare
    b. Coveted
    c. Leadership
    d. Bravery
    e. Honorable

9. ZEAL
    a. Craziness
    b. Resistance
    c. Fervor
    d. Opposition
    e. Apprehension

10. EXTOL
    a. Glorify
    b. Demonize
    c. Chide
    d. Admonish
    e. Criticize

11. REPROACH
    a. Locate
    b. Blame
    c. Concede
    d. Orate
    e. Honor

12. MILIEU
    a. Bacterial
    b. Damp
    c. Ancient
    d. Environment
    e. Uncertain

13. GUILE
    a. Masculine
    b. Stubborn
    c. Naïve
    d. Gullible
    e. Deception

14. ASSENT
   a. Acquiesce
   b. Climb
   c. Assert
   d. Demand
   e. Heighten

15. DEARTH
   a. Grounded
   b. Scarcity
   c. Lethal
   d. Risky
   e. Hearty

16. CONSPICUOUS
   a. Scheme
   b. Obvious
   c. Secretive
   d. Ballistic
   e. Paranoid

17. ONEROUS
   a. Responsible
   b. Generous
   c. Hateful
   d. Burdensome
   e. Wealthy

18. BANAL
   a. Inane
   b. Novel
   c. Painful
   d. Complimentary
   e. Inspired

19. CONTRITE
   a. Tidy
   b. Unrealistic
   c. Contrived
   d. Corrupt
   e. Remorseful

20. MOLLIFY
   a. Pacify
   b. Blend
   c. Negate
   d. Amass
   e. Emote

21. CAPRICIOUS
    a. Skillful
    b. Sanguine
    c. Chaotic
    d. Fickle
    e. Agreeable

22. PALTRY
    a. Appealing
    b. Worthy
    c. Trivial
    d. Fancy
    e. Disgusting

23. SHIRK
    a. Counsel
    b. Evade
    c. Diminish
    d. Sharp
    e. Annoy

24. ASSUAGE
    a. Irritate
    b. Persuade
    c. Argue
    d. Redirect
    e. Soothe

25. TACIT
    a. Unspoken
    b. Shortened
    c. Tenuous
    d. Regal
    e. Timid

Answer Explanations

1. E: Someone who is *wary* is overly cautious or apprehensive. This word is often used in the context of being watchful or on guard about a potential danger. For example, darkening clouds and white caps on the waves may make a seaman wary against setting sail.

2. E: *Proximity* is defined as closeness, or the state or quality of being near in place, time, or relation.

3. C: As an adjective, *parsimonious* means frugal to the point of being stingy, or very unwilling to spend money, which is similar to being miserly.

4. B: The noun *propriety* means suitability or appropriateness to the given circumstances or purpose. It can mean conformity to accepted standards, particularly as they relate to good behavior or manners. In this way, *propriety* can be considered to mean the state or quality of being proper. For example, beyond simply upholding the law, Americans typically expect their president to act with propriety.

5. C: A *boon* is a benefit or blessing, often considered to be timely. It is something to be thankful for. For example, a new tax benefit enacted for first-time homeowners would be a boon to a family who just closed on a house. In an alternative usage, it can be a favor or a benefit given upon request.

6. E: *Strife* is a noun that is defined as bitter or vigorous discord, conflict, or dissension. It can mean a fight or struggle, or other act of contention. For example, antagonistic political interest groups vying for local support may be at strife.

7. B: *Qualm* is a noun that means a feeling of apprehension or uneasiness, often brought on suddenly. A girl who is just learning to ride a bike may have qualms about getting back on the saddle after taking a bad fall. It may also refer to an uneasy feeling related to one's conscience as it pertains to his or her actions. For example, a man with poor morals may have no qualms about lying on his tax return.

8. D: *Valor* is bravery or courage when facing a formidable danger. It often relates to strength of mind or spirit during battle, or acting heroically in such situations.

9. C: *Zeal* is eagerness, fervor, or ardent desire in the pursuit of something. For example, a competitive collegiate baseball player's zeal to succeed in his sport may compromise his academic performance.

10: A: To *extol* is to highly praise, laud, or glorify. People often extol the achievements of their heroes or mentors.

11. B: To *reproach* means to blame, find fault, or severely criticize. It can also be defined as expressing significant disapproval. It is often used in the phrase "beyond reproach" as in, "her violin performance was beyond reproach." In this context, it means her playing was so good that it evaded any possibility of criticism.

12. D: *Milieu* refers to the surroundings or environment.

13. E: *Guile* can be defined as the quality of being cunning or crafty and skilled in deception. Someone may use guile to trick or deceive someone, like to get money from them, or otherwise dupe them.

14. A: As a verb, *assent* means to express agreement, give consent, or acquiesce. A job candidate might assent to an interviewer's request to perform a background check. As a noun, it means an agreement, acceptance, or acquiescence.

15. B: *Dearth* means a lack of something or a scarcity or shortage. For example, a local library might have a dearth of information pertaining to an esoteric topic.

16. B: *Conspicuous* means to be visually or mentally obvious. Something conspicuous stands out, is clearly visible, or may attract attention.

17. D: *Onerous* most closely means burdensome or troublesome. It usually is used to describe a task or obligation that may impose a hardship or burden, often which may be perceived to outweigh its benefits.

18: A: Something *banal* lacks originality, and may be boring and trite. For example, a banal compliment is likely to be a common platitude. Like something that is inane, a banal compliment might be meaningless and lack a convincing quality or significance.

19. E: *Contrite* means to feel or express remorse, or to be regretful and interested in repenting. The noun *contrition* refers to severe remorse or penitence.

20. A: *Mollify* means to sooth, pacify, or appease. It usually is used to refer to reducing the anger or softening the feelings or temper of another person, or otherwise calm them down. For example, a customer service associate may need to mollify an irate customer who is furious about the defect in his or her purchase.

21. D: Of the provided choices, *fickle* is closest in meaning to *capricious*. A person who is capricious tends to display erratic or unpredictable behavior, which is similar to fickle, which is also likely to change spontaneously or behave erratically.

22. C: Although *paltry* often is used as an adjective to describe a very small or meager amount (of money, in particular), it can also mean something trivial or insignificant.

23. B: The word *shirk* means to evade, and is often used in the context of shirking a responsibility, duty, or work.

24. E: *Assuage* most nearly means to soothe or comfort, as in to assuage one's fears. It can also mean to lessen or make less severe, or to relieve. For example, an ice pack on a swollen knee may assuage the pain.

25. A: Something that is *tacit* is usually unspoken but implied. Tacit approval, for example, occurs when agreement or approval is understood without explicitly stating it.

Math Knowledge

Exponents and Roots

The *n*th root of a is given as $\sqrt[n]{a}$, which is called a **radical.** Typical values for *n* are 2 and 3, which represent the square and cube roots. In this form, *n* represents an integer greater than or equal to 2, and a is a real number. If *n* is even, *a* must be nonnegative, and if *n* is odd, *a* can be any real number. This radical can be written in exponential form as $a^{\frac{1}{n}}$. Therefore, $\sqrt[4]{15}$ is the same as $15^{\frac{1}{4}}$ and $\sqrt[3]{-5}$ is the same as $(-5)^{\frac{1}{3}}$.

In a similar fashion, the *n*th root of *a* can be raised to a power *m*, which is written as $\left(\sqrt[n]{a}\right)^{m}$. This expression is the same as $\sqrt[n]{a^m}$. For example:

$$\sqrt[2]{4^3} = \sqrt[2]{64} = 8 = \left(\sqrt[2]{4}\right)^3 = 2^3$$

Because $\sqrt[n]{a} = a^{\frac{1}{n}}$, both sides can be raised to an exponent of *m*, resulting in:

$$\left(\sqrt[n]{a}\right)^{m} = \sqrt[n]{a^m} = a^{\frac{m}{n}}$$

This rule allows:

$$\sqrt[2]{4^3} = \left(\sqrt[2]{4}\right)^3 = 4^{\frac{3}{2}} = (2^2)^{\frac{3}{2}} = 2^{\frac{6}{2}} = 2^3 = 8$$

Negative exponents can also be incorporated into these rules. Any time an exponent is negative, the base expression must be flipped to the other side of the fraction bar and rewritten with a positive exponent. For instance, $2^{-3} = \frac{1}{2^3} = \frac{1}{8}$. Therefore, two more relationships between radical and exponential expressions are:

$$a^{-\frac{1}{n}} = \frac{1}{\sqrt[n]{a}}$$

$$a^{-\frac{m}{n}} = \frac{1}{\sqrt[n]{a^m}} = \frac{1}{\left(\sqrt[n]{a}\right)^{m}}$$

Thus:

$$8^{-3} = \frac{1}{\sqrt[3]{8}} = \frac{1}{2}$$

All of these relationships are very useful when simplifying complicated radical and exponential expressions. If an expression contains both forms, use one of these rules to change the expression to contain either all radicals or all exponential expressions. This process makes the entire expression much easier to work with, especially if the expressions are contained within equations.

Consider the following example: $\sqrt{x} \times \sqrt[4]{x}$. It is written in radical form; however, it can be simplified into one radical by using exponential expressions first. The expression can be written as $x^{\frac{1}{2}} \times x^{\frac{1}{4}}$. It can be combined into one base by adding the exponents as: $x^{\frac{1}{2}+\frac{1}{4}} = x^{\frac{3}{4}}$. Writing this back in radical form, the result is $\sqrt[4]{x^3}$.

Using Structure to Isolate or Identify a Quantity of Interest

When solving equations, it is important to note which quantity must be solved for. This quantity can be referred to as the **quantity of interest**. The goal of solving is to isolate the variable in the equation using logical mathematical steps. The **addition property of equality** states that the same real number can be added to both sides of an equation and equality is maintained. Also, the same real number can be subtracted from both sides of an equation to maintain equality. Second, the **multiplication property of equality** states that the same nonzero real number can multiply both sides of an equation, and still, equality is maintained. Because division is the same as multiplying times a reciprocal, an equation can be divided by the same number on both sides as well.

When solving inequalities, the same ideas are used. However, when multiplying by a negative number on both sides of an inequality, the inequality symbol must be flipped in order to maintain the logic. The same is true when dividing both sides of an inequality by a negative number.

Basically, in order to isolate a quantity of interest in either an equation or inequality, the same thing must be done to both sides of the equals sign, or inequality symbol, to keep everything mathematically correct.

Interpreting the Variables and Constants in Expressions

A linear function of the form $f(x) = mx + b$ has two important quantities: m and b. The quantity m represents the slope of the line, and the quantity b represents the y-intercept of the line. When the function represents an actual real-life situation, or mathematical model, these two quantities are very meaningful. The **slope**, m, represents the rate of change, or the amount y increases or decreases given an increase in x. If m is positive, the rate of change is positive, and if m is negative, the rate of change is negative. The y-intercept, b, represents the amount of the quantity y when x is 0. In many applications, if the x-variable is never a negative quantity, the y-intercept represents the initial amount of the quantity y. Often the x-variable represents time, so it makes sense that the x-variable is never negative.

Consider the following example. These two equations represent the cost, C, of t-shirts, x, at two different printing companies:

$$C(x) = 7x$$

$$C(x) = 5x + 25$$

The first equation represents a scenario that shows the cost per t-shirt is $7. In this equation, x varies directly with y. There is no y-intercept, which means that there is no initial cost for using that printing company. The rate of change is 7, which is price per shirt. The second equation represents a scenario that has both an initial cost and a cost per t-shirt. The slope 5 shows that each shirt is $5. The y-intercept 25 shows that there is an initial cost of using that company. Therefore, it makes sense to use the first company at $7 a shirt when only purchasing a small number of t-shirts. However, any large orders would be cheaper by going with the second company because eventually that initial cost will be negligible.

Recognizing and Representing Patterns

Number and Shape Patterns

Patterns in math are those sets of numbers or shapes that follow a rule. Given a set of values, patterns allow the question of "what's next?" to be answered. In the following set, there are two types of shapes, a white rectangle and a gray circle. The set contains a pattern because every odd-placed shape is a white rectangle and every even-placed spot is taken by a gray circle. This is a pattern because there is a rule of white rectangle, then gray circle, that is followed to find the set.

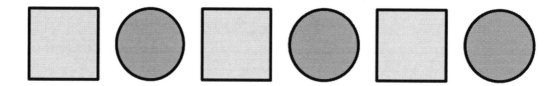

A set of numbers can also be described as having a pattern if there is a rule that can be followed to reproduce the set. The following set of numbers has a rule of adding 3 each time. It begins with zero and increases by 3 each time. By following this rule and pattern, the number after 12 is found to be 15. Further extending the pattern, the number are 18, 21, 24, 27. The pattern of increasing by multiples of three can describe this pattern.

A pattern can also be generated from a given rule. Starting with zero, the rule of adding 5 can be used to produce a set of numbers. The following list will result from using the rule: 0, 5, 10, 15, 20. Describing this pattern can include words such as "multiples" of 5 and an "increase" of 5. Any time this pattern needs to be extended, the rule can be applied to find more numbers. Patterns are identified by the rules they follow. This rule should be able to generate new numbers or shapes, while also applying to the given numbers or shapes.

Making Predictions Based on Patterns

Given a certain pattern, future numbers or shapes can be found. Pascal's triangle is an example of a pattern of numbers. Questions can be asked of the triangle, such as, "what comes next?" and "what values determine the next line?" By examining the different parts of the triangle, conjectures can be made about how the numbers are generated. For the first few rows of numbers, the increase is small. Then the numbers begin to increase more quickly. By looking at each row, a conjecture can be made that the sum of the first row determines the second row's numbers. The second row's numbers can be added to find the third row. To test this conjecture, two numbers can be added, and the number found directly between and below them should be that sum. For the third row, the middle number is 2, which is the sum of the two 1s above it. For the fifth row, the 1 and 3 can be added to find a sum of 4, the same number below the 1 and 3. This pattern continues throughout the triangle. Once the pattern is confirmed to apply throughout the triangle, predictions can be made for future rows. The sums of the bottom row numbers can be found and then added to the bottom of the triangle. In more general terms, the diagonal rows have patterns as well. The outside numbers are always 1. The second diagonal rows are in counting order. The third diagonal row increases each time by one more than the previous. It is helpful to generalize patterns because it makes the pattern more useful in terms of applying it. Pascal's triangle can be used to predict the tossing of a coin, predicting the chances of heads or tails for different numbers of tosses.

It can also be used to show the Fibonacci Sequence, which is found by adding the diagonal numbers together.

Pascal's Triangle

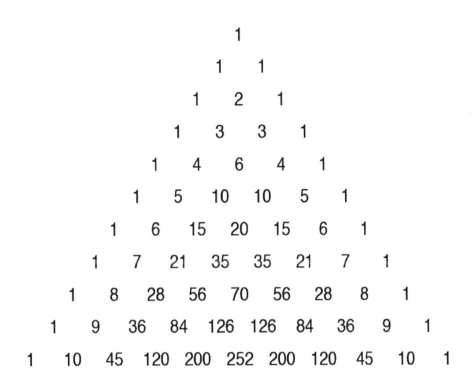

```
                            1
                         1     1
                      1     2     1
                   1     3     3     1
                1     4     6     4     1
             1     5    10    10     5     1
          1     6    15    20    15     6     1
       1     7    21    35    35    21     7     1
    1     8    28    56    70    56    28     8     1
 1     9    36    84   126   126    84    36     9     1
1    10    45   120   200   252   200   120    45    10     1
```

Relationships Between the Corresponding Terms of Two Numerical Patterns

Sets of numerical patterns can be found by starting with a number and following a given rule. If two sets are generated, the corresponding terms in each set can be found to relate to one another by one or more operations. For example, the following table shows two sets of numbers that each follow their own pattern. The first column shows a pattern of numbers increasing by 1. The second column shows the numbers increasing by 4. Because the numbers are lined up, corresponding numbers are side by side for the two sets. A question to ask is, "How can the number in the first column be turned into the number in the second column?"

1	4
2	8
3	12
4	16
5	20

This answer will lead to the relationship between the two sets. By recognizing the multiples of 4 in the right column and the counting numbers in order in the left column, the relationship of multiplying by four is determined. The first set is multiplied by 4 to get the second set of numbers. To confirm this relationship, each set of corresponding numbers can be checked. For any two sets of numerical patterns, the corresponding numbers can be lined up to find how each one relates to the other. In some cases, the relationship is simply addition or subtraction, multiplication or division. In other relationships, these operations are used in conjunction with each together. As seen in the following table, the relationship uses multiplication and addition. The following expression shows this relationship: $3x + 2$. The x represents the numbers in the first column.

1	5
2	8
3	11
4	14

Algebra

Algebraic Expressions Versus Equations

An **algebraic expression** is a mathematical phrase that may contain numbers, variables, and mathematical operations. An expression represents a single quantity. For example, $3x + 2$ is an algebraic expression.

An **algebraic equation** is a mathematical sentence with two expressions that are equal to each other. That is, an equation must contain an equals sign, as in $3x + 2 = 17$. This statement says that the value of the expression on the left side of the equals sign is equivalent to the value of the expression on the right side. In an expression, there are not two sides because there is no equals sign. The equals sign (=) is the difference between an expression and an equation.

To distinguish an expression from an equation, just look for the equals sign.

Example: Determine whether each of these is an expression or an equation.

    a. $16 + 4x = 9x - 7$ Solution: Equation

    b. $-27x - 42 + 19y$ Solution: Expression

    c. $4 = x + 3$ Solution: Equation

Adding and Subtracting Linear Algebraic Expressions

To add and subtract linear algebra expressions, you must combine like terms. **Like terms** are described as those terms that have the same variable with the same exponent. In the following example, the x-terms can be added because the variable is the same and the exponent on the variable of one is also the same. These terms add to be $9x$. The other like terms are called **constants** because they have no variable component. These terms will add to be nine.

Example: Add $(3x - 5) + (6x + 14)$

        $3x - 5 + 6x + 14$ Rewrite without parentheses

        $3x + 6x - 5 + 14$ Use the commutative property of addition

        $9x + 9$ Combine like terms

When subtracting linear expressions, be careful to add the opposite when combining like terms. Do this by distributing -1, which is multiplying each term inside the second parenthesis by negative one. Remember that distributing -1 changes the sign of each term.

Example: Subtract $(17x + 3) - (27x - 8)$

$17x + 3 - 27x + 8$	Use the distributive Property
$17x - 27x + 3 + 8$	Use the commutative property of addition
$-10x + 11$	Combine like terms

Example: Simplify by adding or subtracting:

$(6m + 28z - 9) + (14m + 13) - (-4z + 8m + 12)$

$6m + 28z - 9 + 14m + 13 + 4z - 8m - 12$	Use the distributive Property
$6m + 14m - 8m + 28z + 4z - 9 + 13 - 12$	Use the commutative property of addition
$12m + 32z - 8$	Combine like terms

Using the Distributive Property to Generate Equivalent Linear Algebraic Expressions

The Distributive Property: $a(b + c) = ab + ac$

The **distributive property** is a way of taking a factor and multiplying it through a given expression in parentheses. Each term inside the parentheses is multiplied by the outside factor, eliminating the parentheses. The following example shows how to distribute the number 3 to all the terms inside the parentheses.

Example: Use the distributive property to write an equivalent algebraic expression:

$3(2x + 7y + 6)$	
$3(2x) + 3(7y) + 3(6)$	Use the distributive property
$6x + 21y + 18$	Simplify

Because $a - b$ can be written $a + (-b)$, the distributive property can be applied in the example below.

Example: Use the distributive property to write an equivalent algebraic expression.

$7(5m - 8)$	
$7[5m + (-8)]$	Rewrite subtraction as addition of -8
$7(5m) + 7(-8)$	Use the distributive property
$35m - 56$	Simplify

In the following example, note that the factor of 2 is written to the right of the parentheses but is still distributed as before:

Example: Use the distributive property to write an equivalent algebraic expression:

$(3m + 4x - 10)2$

$(3m)2 + (4x)2 + (-10)2$ Use the distributive property

$6m + 8x - 20$ Simplify

Example: $-(-2m + 6x)$

In this example, the negative sign in front of the parentheses can be interpreted as $-1(-2m + 6x)$

$1(-2m + 6x)$

$-1(-2m) + (-1)(6x)$ Use distributive property

$2m - 6x$ Simplify

Evaluating Simple Algebraic Expressions for Given Values of Variables

To evaluate an algebra expression for a given value of a variable, replace the variable with the given value. Then perform the given operations to simplify the expression.

Example: Evaluate $12 + x$ for $x = 9$

$12 + (9)$ Replace x with the value of 9 as given in the problem. *It is a good idea to always use parentheses when substituting this value. This will be particularly important in the following examples.*

21 Add

Now see that when x is 9, the value of the given expression is 21.

Example: Evaluate $4x + 7$ for $x = 3$

$4(3) + 7$ Replace the x in the expression with 3

$12 + 7$ Multiply (remember order of operations)

19 Add

Therefore, when x is 3, the value of the given expression is 19.

Example: Evaluate $-7m - 3r - 18$ for $m = 2$ and $r = -1$

$-7(2) - 3(-1) - 18$ Replace m with 2 and r with -1

$-14 + 3 - 18$ Multiply

-29 Add

So, when m is 2 and r is –1, the value of the given expression is –29.

Mathematical Terms That Identify Parts of Expressions

A **variable** is a symbol used to represent a number. Letters, like x, y, and z are often used as variables in algebra.

A **constant** is a number that cannot change its value. For example, 18 is a constant.

A **term** is a constant, variable, or the product of constants and variables. In an expression, terms are separated by + and − signs. Examples of terms are $24x$, –32, and $15xyz$.

Like terms are terms that contain the same variables in the same powers. For example, $6z$ and $-8z$ are like terms, and $9xy$ and $17xy$ are like terms. $12y^3$ and $3y^3$ are also like terms. Lastly, constants, like 23 and 51, are like terms as well.

A **factor** is something that is multiplied by something else. A factor may be a constant, a variable, or a sum of constants or variables.

A **coefficient** is the numerical factor in a term that has a variable. In the term $16x$, the coefficient is 16.

Example: Given the expression, $6x - 12y + 18$, answer the following questions.

1. How many terms are in the expression?

   Solution: 3

2. Name the terms.

   Solution: 6x, –12y, and 18 (Notice that the minus sign preceding the 12 is interpreted to represent negative 12)

3. Name the factors.

   Solution: 6, x, –12, y

4. What are the coefficients in this expression?

   Solution: 6 and –12

5. What is the constant in this expression?

   Solution: 18

Using Formulas to Determine Unknown Quantities

Given the formula for the area of a rectangle, $A = lw$, with A = area, l = length, and w = width, the area of a rectangle can be determined, given the length and the width.

For example, if the length of a rectangle is 7 cm and the width is 10 cm, find the area of the rectangle. Just as when evaluating expressions, to solve, replace the variables with the given values. Thus, given $A = lw$, and $l = 7$ and $w = 10$, $A = (7)(10)$, which equals 70. Therefore, the area of the rectangle is 70 cm^2.

Consider an example using the formula for perimeter of a rectangle, which is $P = 2l + 2w$, where P is perimeter, l is length, and w is width. If the length of a rectangle is 12 inches and the width is 9 inches, find the perimeter.

Solution: $P = 2l + 2w$

$P = 2(12) + 2(9)$ Replace l with 12 and w with 9

$P = 24 + 18$ Use correct order of operations; multiply first

$P = 42$ Add

The perimeter of this rectangle is 42 inches.

Solving Real-World One- or Multi-Step Problems with Rational Numbers

One-step problems take only one mathematical step to solve. For example, solving the equation $5x = 45$ is a one-step problem because the one step of dividing both sides of the equation by 5 is the only step necessary to obtain the solution $x = 9$. The multiplication principle of equality is the one step used to isolate the variable. The equation is of the form $ax = b$, where a and b are rational numbers. Similarly, the addition principle of equality could be the one step needed to solve a problem. In this case, the equation would be of the form $x + a = b$ or $x - a = b$, for real numbers a and b.

A multi-step problem involves more than one step to find the solution, or it could consist of solving more than one equation. An equation that involves both the addition principle and the multiplication principle is a two-step problem, and an example of such an equation is

$$2x - 4 = 5$$

Solving involves adding 4 to both sides and then dividing both sides by 2. An example of a two-step problem involving two separate equations is

$$y = 3x, 2x + y = 4$$

The two equations form a system of two equations that must be solved together in two variables. The system can be solved by the substitution method. Since y is already solved for in terms of x, plug $3x$ in for y into the equation $2x + y = 4$, resulting in $2x + 3x = 4$. Therefore, $5x = 4$ and $x = \frac{4}{5}$. Because there are two variables, the solution consists of both a value for x and for y. Substitute $x = \frac{4}{5}$ into either original equation to find y. The easiest choice is $y = 3x$. Therefore, $y = 3 \times \frac{4}{5} = \frac{12}{5}$. The solution can be written as the ordered pair $\left(\frac{4}{5}, \frac{12}{5}\right)$.

Real-world problems can be translated into both one-step and multi-step problems. In either case, the word problem must be translated from the verbal form into mathematical expressions and equations that can be solved using algebra. An example of a one-step real-world problem is the following: A cat weighs half as much as a dog living in the same house. If the dog weighs 14.5 pounds, how much does the cat weigh? To solve this problem, an equation can be used. In any word problem, the first step is to define variables that represent the unknown quantities. For this problem, let x be equal to the unknown weight of the cat. Because two times the weight of the cat equals 14.5 pounds, the equation to be solved is: $2x = 14.5$. Use the multiplication principle to divide both sides by 2. Therefore, $x = 7.25$. The cat weighs 7.25 pounds.

Most of the time, real-world problems are more difficult than this one and consist of multi-step problems. The following is an example of a multi-step problem: The sum of two consecutive page numbers is equal to 437. What are those page numbers? First, define the unknown quantities. If x is equal to the first page number, then $x + 1$ is equal to the next page number because they are consecutive integers. Their sum is equal to 437, and this statement translates to the equation $x + x + 1 = 437$. To solve, first collect like terms to obtain $2x + 1 = 437$. Then, subtract 1 from both sides and then divide by 2. The solution to the equation is $x = 218$. Therefore, the two consecutive page numbers that satisfy the problem are 218 and 219. It is always important to make sure that answers to real-world problems make sense. For instance, if the solution to this same problem resulted in decimals, that should be a red flag indicating the need to check the work. Page numbers are whole numbers; therefore, if decimals are found to be answers, the solution process should be double-checked to see where mistakes were made.

Solving Equations in One Variable

An **equation in one variable** is a mathematical statement where two algebraic expressions in one variable, usually x, are set equal. To solve the equation, the variable must be isolated on one side of the equals sign. The addition and multiplication principles of equality are used to isolate the variable. The **addition principle of equality** states that the same number can be added to or subtracted from both sides of an equation. Because the same value is being used on both sides of the equals sign, equality is maintained. For example, the equation $2x - 3 = 5x$ is equivalent to both:

$$(2x - 3) + 3 = 5x + 3$$

and

$$(2x - 3) - 5 = 5x - 5$$

This principle can be used to solve the following equation: $x + 5 = 4$. The variable x must be isolated, so to move the 5 from the left side, subtract 5 from both sides of the equals sign. Therefore, $x + 5 - 5 = 4 - 5$. So, the solution is $x = -1$. This process illustrates the idea of an **additive inverse** because subtracting 5 is the same as adding -5. Basically, add the opposite of the number that must be removed to both sides of the equals sign.

The **multiplication principle of equality** states that equality is maintained when a number is either multiplied times both expressions on each side of the equals sign, or when both expressions are divided by the same number. For example, $4x = 5$ is equivalent to both $16x = 20$ and $x = \frac{5}{4}$. Multiplying both sides times 4 and dividing both sides by 4 maintains equality. Solving the equation $6x - 18 = 5$ requires the use of both principles. First, apply the addition principle to add 18 to both sides of the equals sign, which results in $6x = 23$. Then use the multiplication principle to divide both sides by 6, giving the solution $x = \frac{23}{6}$. Using the multiplication principle in the solving process is the same as involving a multiplicative inverse. A **multiplicative inverse** is a value that, when multiplied by a given number, results in 1. Dividing by 6 is the same as multiplying by $\frac{1}{6}$, which is both the reciprocal and multiplicative inverse of 6.

When solving a linear equation in one variable, checking the answer shows if the solution process was performed correctly. Plug the solution into the variable in the original equation. If the result is a false statement, something was done incorrectly during the solution procedure. Checking the example above gives the following:

$$6 \times \frac{23}{6} - 18 = 23 - 18 = 5$$

Therefore, the solution is correct.

Some equations in one variable involve fractions or the use of the distributive property. In either case, the goal is to obtain only one variable term and then use the addition and multiplication principles to isolate that variable. Consider the equation $\frac{2}{3}x = 6$. To solve for x, multiply each side of the equation by the reciprocal of $\frac{2}{3}$, which is $\frac{3}{2}$. This step results in $\frac{3}{2} \times \frac{2}{3}x = \frac{3}{2} \times 6$, which simplifies into the solution $x = 9$. Now consider the equation $3(x + 2) - 5x = 4x + 1$. Use the distributive property to clear the parentheses. Therefore, multiply each term inside the parentheses by 3. This step results in $3x + 6 - 5x = 4x + 1$. Next, collect like terms on the left-hand side. **Like terms** are terms with the same variable or variables raised to the same exponent(s). Only like terms can be combined through addition or subtraction. After collecting like terms, the equation is $-2x + 6 = 4x + 1$. Finally, apply the addition and multiplication principles. Add $2x$ to both sides to obtain $6 = 6x + 1$. Then, subtract 1 from both sides to obtain $5 = 6x$. Finally, divide both sides by 6 to obtain the solution $\frac{5}{6} = x$.

Two other types of solutions can be obtained when solving an equation in one variable. The final result could be that there is either no solution or that the solution set contains all real numbers. Consider the equation $4x = 6x + 5 - 2x$. First, the like terms can be combined on the right to obtain $4x = 4x + 5$. Next, subtract $4x$ from both sides. This step results in the false statement $0 = 5$. There is no value that can be plugged into x that will ever make this equation true. Therefore, there is no solution. The solution procedure contained correct steps, but the result of a false statement means that no value satisfies the equation. The symbolic way to denote that no solution exists is ∅. Next, consider the equation $5x + 4 + 2x = 9 + 7x - 5$. Combining the like terms on both sides results in $7x + 4 = 7x + 4$. The left-hand side is exactly the same as the right-hand side. Using the addition principle to move terms, the result is $0 = 0$, which is always true. Therefore, the original equation is true for any number, and the solution set is all real numbers. The symbolic way to denote such a solution set is ℝ, or in interval notation, $(-\infty, \infty)$.

Solving a Linear Inequality in One Variable

A **linear equation** in x can be written in the form $ax + b = 0$. A **linear inequality** is very similar, although the equals sign is replaced by an inequality symbol such as $<, >, \leq$, or $\geq$. In any case, a can never be 0. Some examples of linear inequalities in one variable are $2x + 3 < 0$ and $4x - 2 \leq 0$. Solving an inequality involves finding the set of numbers that when plugged into the variable, make the inequality a true statement. These numbers are known as the **solution set** of the inequality. To solve an inequality, use the same properties that are necessary in solving equations. First, add or subtract variable terms and/or constants to obtain all variable terms on one side of the equals sign and all constant terms on the other side. Then, either multiply both sides times the same number, or divide both sides by the same number, to obtain an inequality that gives the solution set. When multiplying times, or dividing by, a negative number in an inequality, change the direction of the inequality symbol. The solution set can be graphed on a number line. Consider the linear inequality $-2x - 5 > x + 6$. First, add 5 to both sides and subtract $-x$ off of both sides to obtain $-3x > 11$. Then, divide both sides by -3, making sure to change the direction of the inequality symbol. These steps result in the solution $x < -\frac{11}{3}$. Therefore, any number less than $-\frac{11}{3}$ satisfies this inequality.

Translating Phrases and Sentences into Expressions, Equations, and Inequalities

When presented with a real-world problem that must be solved, the first step is always to determine what the unknown quantity is that must be solved for. Use a variable, such as x or t, to represent that unknown quantity. Sometimes, there can be two or more unknown quantities. In this case, either choose an additional variable, or if a relationship exists between the unknown quantities, express the other quantities in terms of the original variable. After choosing the variables, form algebraic expressions and/or equations that represent the verbal statement in the problem. The following table shows examples of vocabulary used to represent the different operations:

Addition	Sum, plus, total, increase, more than, combined, in all
Subtraction	Difference, less than, subtract, reduce, decrease, fewer, remain
Multiplication	Product, multiply, times, part of, twice, triple
Division	Quotient, divide, split, each, equal parts, per, average, shared

The combination of operations and variables form both mathematical expression and equations. As mentioned, the difference between expressions and equations are that there is no equals sign in an expression, and that expressions are **evaluated** to find an unknown quantity, while equations are **solved** to find an unknown quantity. Also, inequalities can exist within verbal mathematical statements. Instead of a statement of equality, expressions state quantities are *less than, less than or equal to, greater than,* or *greater than or equal to.* Another type of inequality is when a quantity is said to be *not equal to* another quantity. The symbol used to represent "not equal to" is $\neq$.

The steps for solving inequalities in one variable are the same steps for solving equations in one variable. The addition and multiplication principles are used. However, to maintain a true statement when using the $<$, $\leq$, $>$, and $\geq$ symbols, if a negative number is either multiplied times both sides of an inequality or divided from both sides of an inequality, the sign must be flipped. For instance, consider the following inequality: $3 - 5x \leq 8$. First, 3 is subtracted from each side to obtain $-5x \leq 5$. Then, both sides are divided by -5, while flipping the sign, to obtain $x \geq -1$. Therefore, any real number greater than or equal to -1 satisfies the original inequality.

Rewriting Simple Rational Expressions

A **rational expression** is a fraction or a ratio in which both the numerator and denominator are polynomials that are not equal to zero. A *polynomial* is a mathematical expression containing the sum and difference of one or more terms that are constants multiplied times variables raised to positive powers. Here are some examples of rational expressions:

$$\frac{2x^2 + 6x}{x}$$

$$\frac{x - 2}{x^2 - 6x + 8}$$

$$\frac{x + 2}{x^3 - 1}$$

Such expressions can be simplified using different forms of division. The first example can be simplified in two ways. First, because the denominator is a monomial, the expression can be split up into two expressions: $\frac{2x^2}{x} + \frac{6x}{x}$, and then simplified using properties of exponents as $2x + 6$. It also can be simplified using factoring and then crossing common factors out of the numerator and denominator. For instance, it can be written as:

$$\frac{2x(x+3)}{x} = 2(x+3) = 2x + 6$$

The second expression above can also be simplified using factoring. It can be written as:

$$\frac{x-2}{(x-2)(x-4)} = \frac{1}{x-4}$$

Finally, the third example can only be simplified using long division, as there are no common factors in the numerator and denominator. First, divide the first term of the denominator by the first term of the numerator, then write that in the quotient. Then, multiply the divisor times that number and write it below the dividend. Subtract, bring down the next term from the dividend, and continue that process with the next first term and first term of the divisor. Continue the process until every term in the divisor is accounted for. Here is the actual long division:

Simplifying Expressions Using Long Division

$$
\begin{array}{r}
x^2 - 2x + 4 \\
x+2 \enclose{longdiv}{x^3 - 1} \\
\underline{x^3 + 2x^2} \\
-2x^2 - 1 \\
\underline{-2x^2 - 4x} \\
4x - 1 \\
\underline{4x + 8} \\
-9
\end{array}
$$

Function Notation

A **relation** is any set of ordered pairs (x, y). The first set of points, known as the x-coordinates, make up the domain of the relation. The second set of points, known as the y-coordinates, make up the range of the relation. A relation in which every member of the domain corresponds to only one member of the range is known as a **function.** A function cannot have a member of the domain corresponding to two members of the range. Functions are most often given in terms of equations instead of ordered pairs. For instance, here is an equation of a line: $y = 2x + 4$. In function notation, this can be written as

$$f(x) = 2x + 4$$

The expression $f(x)$ is read "f of x" and it shows that the inputs, the x-values, get plugged into the function and the output is $y = f(x)$. The set of all inputs are in the domain and the set of all outputs are in the range.

The x-values are known as the **independent variables** of the function and the y-values are known as the **dependent variables** of the function. The y-values depend on the x-values. For instance, if $x = 2$ is plugged into the function shown above, the y-value depends on that input.

$$f(2) = 2 \times 2 + 4 = 8.$$

Therefore, $f(2) = 8$, which is the same as writing the ordered pair (2, 8). To graph a function, graph it in equation form and plot ordered pairs.

Due to the definition of a function, the graph of a function cannot have two of the same x-components paired to different y-component. For example, the ordered pairs (3, 4) and (3, -1) cannot be in a valid function. Therefore, all graphs of functions pass the **vertical line test**. If any vertical line intersects a graph in more than one place, the graph is not that of a function. For instance, the graph of a circle is not a function because one can draw a vertical line through a circle and the would intersect the circle twice. Common functions include lines and polynomials, and they all pass the vertical line test.

Linear Functions that Model a Linear Relationship

A **linear function that models a linear relationship between two quantities** is of the form $y = mx + b$, or in function form $f(x) = mx + b$. In a linear function, the value of y depends on the value of x, and y increases or decreases at a constant rate as x increases. Therefore, the independent variable is x, and the dependent variable is y. The graph of a linear function is a line, and the constant rate can be seen by looking at the steepness, or slope, of the line. If the line increases from left to right, the slope is positive. If the line slopes downward from left to right, the slope is negative. In the function, m represents slope. Each point on the line is an **ordered pair** (x, y), where x represents the x-coordinate of the point and y represents the y-coordinate of the point. The point where $x = 0$ is known as the y-intercept, and it is the place where the line crosses the y-axis. If $x = 0$ is plugged into $f(x) = mx + b$, the result is $f(0) = b$, so therefore, the point $(0, b)$ is the y-intercept of the line. The derivative of a linear function is its slope.

Consider the following situation. A taxicab driver charges a flat fee of $2 per ride and $3 a mile. This statement can be modeled by the function $f(x) = 3x + 2$ where x represents the number of miles and $f(x) = y$ represents the total cost of the ride. The total cost increases at a constant rate of $2 per mile, and that is why this situation is a linear relationship. The slope $m = 3$ is equivalent to this rate of change. The flat fee of $2 is the y-intercept. It is the place where the graph crosses the x-axis, and it represents the cost when $x = 0$, or when no miles have been traveled in the cab. The y-intercept in this situation represents the flat fee.

Polynomial Functions

A **polynomial function** is a function containing a polynomial expression, which is an expression containing constants and variables combined using the four mathematical operations. The degree of a polynomial in one variable is the largest exponent seen on any variable in the expression. Typical polynomial functions are **quartic,** with a degree of 4, **cubic,** with a degree of 3, and **quadratic,** with a degree of 2. Note that the exponents on the variables can only be nonnegative integers. The domain of any polynomial function is all real numbers because any number plugged into a polynomial expression grants a real number output. An example of a quartic polynomial equation is $y = x^4 + 3x^3 - 2x + 1$. The zeros of a polynomial function are the points where its graph crosses the y-axis. In order to find the number of real zeros of a polynomial function, **Descartes' Rule of Sign** can be used. The number of possible positive real zeros is equal to the number of sign changes in the coefficients of the terms in the polynomial. If there is only one sign change, there is only one positive real zero. In the example above, the signs of the coefficients are positive, positive, negative, and positive. Therefore, the sign changes two times and therefore, there are at most two positive real zeros. The number of possible negative real zeros is equal to the number of sign changes in the coefficients when plugging $-x$ into the equation. Again, if there is only one sign change, there is only one negative real zero. The polynomial result when plugging -x into the equation is

$$y^4 - 3x^3 + 2x + 1$$

The sign changes two times, so there are at most two negative real zeros. Another polynomial equation this rule can be applied to is

$$y = x^3 + 2x - x - 5$$

There is only one sign change in the terms of the polynomial, so there is exactly one real zero. When plugging -x into the equation, the polynomial result is

$$-x^3 - 2x - x - 5$$

There are no sign changes in this polynomial, so there are no possible negative zeros.

Adding, Subtracting, and Multiplying Polynomial Equations

When working with polynomials, **like terms** are terms that contain exactly the same variables with the same powers. For example, $x^4 y^5$ and $9x^4 y^5$ are like terms. The coefficients are different, but the same variables are raised to the same powers. When adding polynomials, only terms that are like can be added. When adding two like terms, just add the coefficients and leave the variables alone. This process uses the distributive property. For example:

$$x^4 y^5 + 9x^4 y^5 = (1 + 9)x^4 y^5 = 10x^4 y^5$$

Therefore, when adding two polynomials, simply add the like terms together. Unlike terms cannot be combined.

Subtracting polynomials involves adding the opposite of the polynomial being subtracted. Basically, the sign of each term in the polynomial being subtracted is changed, and then the like terms are combined because it is now an addition problem. For example, consider the following:

$$6x^2 - 4x + 2 - (4x^2 - 8x + 1).$$

Add the opposite of the second polynomial to obtain $6x^2 - 4x + 2 + (-4x^2 + 8x - 1)$. Then, collect like terms to obtain $2x^2 + 4x + 1$.

Multiplying polynomials involves using the product rule for exponents that $b^m b^n = b^{m+n}$. Basically, when multiplying expressions with the same base, just add the exponents. Multiplying a monomial by a monomial involves multiplying the coefficients together and then multiplying the variables together using the product rule for exponents. For instance, $8x^2y \times 4x^4y^2 = 32x^6y^3$. When multiplying a monomial by a polynomial that is not a monomial, use the distributive property to multiply each term of the polynomial times the monomial. For example, $3x(x^2 + 3x - 4) = 3x^3 + 9x^2 - 12x$. Finally, multiplying two polynomials when neither one is a monomial involves multiplying each term of the first polynomial times each term of the second polynomial. There are some shortcuts, given certain scenarios. For instance, a binomial times a binomial can be found by using the **FOIL (Firsts, Outers, Inners, Lasts)** method shown here.

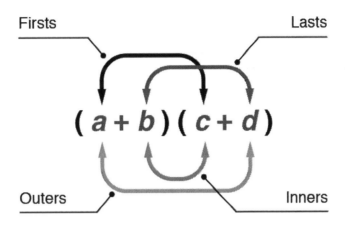

Finding the product of a sum and difference of the same two terms is simple because if it was to be foiled out, the outer and inner terms would cancel out. For instance, $(x + y)(x - y) = x^2 + xy - xy - y^2$. Finally, the square of a binomial can be found using the following formula: $(a \pm b)^2 = a^2 \pm 2ab + b^2$.

The Relationship Between Zeros and Factors of Polynomials

A **polynomial** is a mathematical expression containing the sum and difference of one or more terms that are constants multiplied times variables raised to positive powers. A **polynomial equation** is a polynomial set equal to another polynomial, or in standard form, a polynomial is set equal to 0. A **polynomial function** is a polynomial set equal to y. For instance, $x^2 + 2x - 8$ is a polynomial, $x^2 + 2x - 8 = 0$ is a polynomial equation, and $y = x^2 + 2x - 8$ is the corresponding polynomial function. To solve a polynomial equation, the x-values in which the graph of the corresponding polynomial function crosses the x-axis are sought. These coordinates are known as the **zeros** of the polynomial function, because they are the coordinates in which the y-coordinates are 0. One way to find the zeros of a polynomial is to find its factors, then set each individual factor equal to 0, and solve each equation to find the zeros. A **factor** is a linear expression, and to completely factor a polynomial, the polynomial must be rewritten as a product of individual linear factors. The polynomial listed above can be factored as $(x + 4)(x - 2)$. Setting each factor equal to zero results in the zeros $x = -4$ and $x = 2$.

Here is the graph of the zeros of the polynomial:

The Graph of the Zeros of x² + 2x - 8 = 0

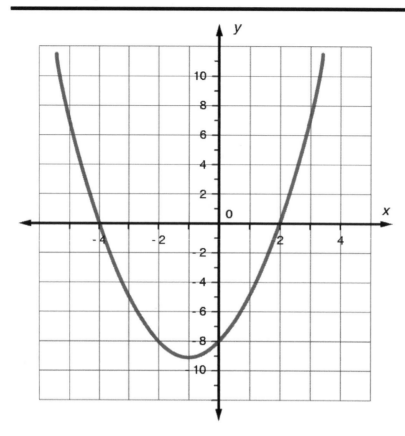

Interpreting Solutions of Multistep One-Variable Linear Equations and Inequalities

Multistep one-variable equations involve the use of one variable in an equation with many operations. For example, the equation $2x + 4 = 10$ involves one variable, x, and multiple steps to solve for the value of x. The first step is the move the four to the opposite side of the equation by subtracting 4. The next step is to divide by 2. The final answer yields a value of 3 for the variable x. The steps for this process are shown below:

$$2x + 4 = 10$$

$$ -4 \quad -4 \qquad \text{Subtract 4 on both sides}$$

$$2x = 6$$

$$\div 2 \quad \div 2 \qquad \text{Divide by 2 on both sides}$$

$$x = 3$$

When the result is found, the value of the variable must be interpreted. For this problem, a value of 3 can be understood as the amount that can be doubled and then increased by 4 to yield a value of 10.

Inequalities can also be interpreted in much the same way. The following inequality can be solved to find the value of

$$b. \frac{b}{7} - 8 \geq 7$$

This inequality models the amount of money a group of friends earned for cleaning up a neighbor's yard, b. There were 7 friends, so the money had to be split seven times. Then $8 was taken away from each friend to pay for materials they bought to help clean the yard. All these things needed to be less than or equal to seven for the friends to each receive at least $7. The first step is to add 8 to both sides of the inequality. Then the 7 can be multiplied on each side. The resulting inequality is $b \geq 105$. Because the answer is not only an equals sign, the value for b is not a single number. In this problem, the answer communicates that the value of b must be greater than or equal to $105 in order for each friend to make at least $7 for their work. The number for b, what they are paid, can be more than 105 because that would mean they earned more money. They do not want it to be less than 105 because their profit will drop below $7 per piece.

Linear Relationships Represented by Graphs, Equations, and Tables

Graphs, equations, and tables are three different ways to represent linear relationships. The following graph shows a linear relationship because the relationship between the two variables is constant. As the distance increases by 25 miles, the time lapses by 1 hour. This pattern continues for the rest of the graph. The line represents a constant rate of 25 miles per hour. This graph can also be used to solve problems involving predictions for a future time. After 8 hours of travel, the rate can be used to predict the distance covered. Eight hours of travel at 25 miles per hour covers a distance of 200 miles. The equation at the top of the graph corresponds to this rate also. The same prediction of distance in a given time can be found using the equation.

For a time of 10 hours, the distance would be 250 miles, as the equation yields $d = 25 \times 10 = 250$.

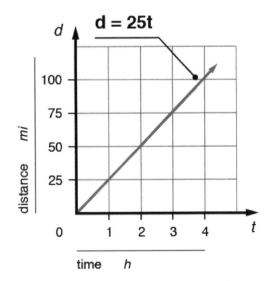

Another representation of a linear relationship can be seen in a table. The first thing to observe from the table is that the y-values increase by the same amount of 3 each time. As the x-values increase by 1, the y-values increase by 3. This pattern shows that the relationship is linear. If this table shows the money earned, y-value, for the hours worked, x-value, then it can be used to predict how much money will be earned for future hours. If 6 hours are worked, then the pay would be $19. For further hours and money to be determined, it would be helpful to have an equation that models this table of values. The equation will show the relationship between x and y. The y-value can each time be determined by multiplying the x-value by 3, then adding 1. The following equation models this relationship: $y = 3x + 1$. Now that there is an equation, any number of hours, x, can be substituted into the equation to find the amount of money earned, y.

y = 3x + 1	
x	y
0	1
1	4
2	7
4	13
5	16

Graphing and Statistics

The Coordinate Plane

The coordinate plane is a way of identifying the position of a point in relation to two axes. The **coordinate plane** is made up of two intersecting lines, the *x*- and *y*-axes. These lines intersect at a right angle, and their intersection point is called the **origin.** The points on the coordinate plane are labeled based on their position in relation to the origin. If a point is found 4 units to the right and 2 units up from the origin, the location is described as (4, 2). These numbers are the *x*- and *y*-coordinates, always written in the order (*x*, *y*). This point is also described as lying in the first quadrant. Every point in the first quadrant has a location that is positive in the *x* and *y* directions. The following figure shows the coordinate plane with examples of points that lie in each quadrant:

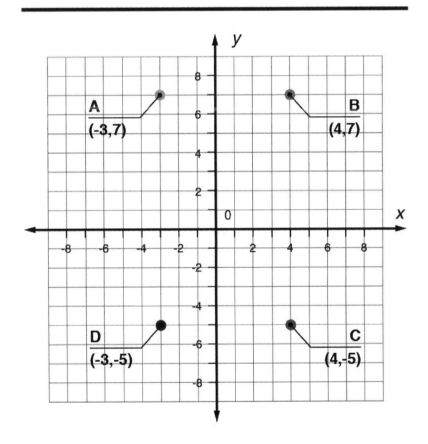

The Coordinate Plane

Point B lies in the first quadrant, described positive *x*- and *y*-values, above the *x*-axis and to the right of the *y*-axis. Point A lies in the second quadrant, where the *x*-value is negative and *y*-value is positive. This quadrant is above the *x*-axis and to the left of the *y*-axis. Point D lies in the third quadrant, where both the *x*- and *y*-values are negative. Below the *x*-axis, and to the left of the *y*-axis, is how this quadrant is described. Point C is in the fourth quadrant, where the *x*-value is positive and the *y*-value is negative.

Tables, Charts, and Graphs

Tables, charts, and graphs can be used to convey information about different variables. They are all used to organize, categorize, and compare data, and they all come in different shapes and sizes. Each type has its own way of showing information, whether it is in a column, shape, or picture. To answer a question relating to a table, chart, or graph, some steps should be followed. First, the problem should be read thoroughly to determine what is being asked to determine what quantity is unknown. Then, the title of the table, chart, or graph should be read. The title should clarify what actual data is being summarized in the table. Next, look at the key and both the horizontal and vertical axis labels, if they are given. These items will provide information about how the data is organized. Finally, look to see if there is any more labeling inside the table. Taking the time to get a good idea of what the table is summarizing will be helpful as it is used to interpret information.

Tables are a good way of showing a lot of information in a small space. The information in a table is organized in columns and rows. For example, a table may be used to show the number of votes each candidate received in an election. By interpreting the table, one may observe which candidate won the election and which candidates came in second and third. In using a bar chart to display monthly rainfall amounts in different countries, rainfall can be compared between countries at different times of the year. Graphs are also a useful way to show change in variables over time, as in a line graph, or percentages of a whole, as in a pie graph.

The table below relates the number of items to the total cost. The table shows that 1 item costs $5. By looking at the table further, 5 items cost $25, 10 items cost $50, and 50 items cost $250. This cost can be extended for any number of items. Since 1 item costs $5, then 2 items would cost $10. Though this information isn't in the table, the given price can be used to calculate unknown information.

Number of Items	1	5	10	50
Cost ($)	5	25	50	250

A **bar graph** is a graph that summarizes data using bars of different heights. It is useful when comparing two or more items or when seeing how a quantity changes over time. It has both a horizontal and vertical axis. Interpreting bar graphs includes recognizing what each bar represents and connecting that to the two variables. The bar graph below shows the scores for six people on three different games. The color of the bar shows which game each person played, and the height of the bar indicates their score for that game. William scored 25 on game 3, and Abigail scored 38 on game 3. By comparing the bars, it's obvious that Williams scored lower than Abigail.

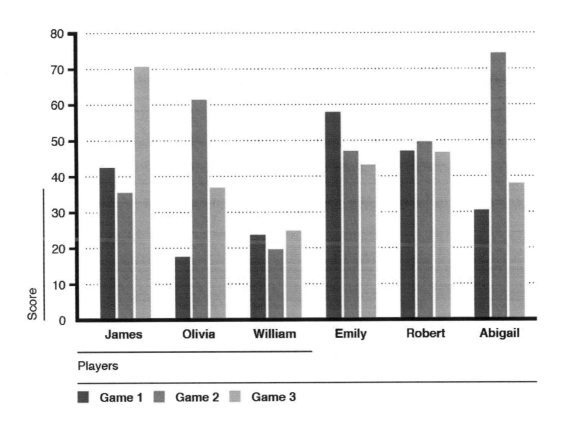

A **line graph** is a way to compare two variables. Each variable is plotted along an axis, and the graph contains both a horizontal and a vertical axis. On a line graph, the line indicates a continuous change. The change can be seen in how the line rises or falls, known as its slope, or rate of change. Often, in line graphs, the horizontal axis represents a variable of time. Audiences can quickly see if an amount has increased or decreased over time. The bottom of the graph, or the *x*-axis, shows the units for time, such as days, hours, months, etc. If there are multiple lines, a comparison can be made between what the two lines represent. For example, the following line graph shows the change in temperature over five days. The top line represents the high, and the bottom line represents the low for each day. Looking at the top line alone, the high decreases for a day, then increases on Wednesday. Then it decreased on Thursday and increases again on Friday. The low temperatures have a similar trend, shown in bottom line. The range in temperatures each day can also be calculated by finding the difference between the top line and bottom line on a particular day. On Wednesday, the range was 14 degrees, from 62 to 76° F.

Daily Temperatures

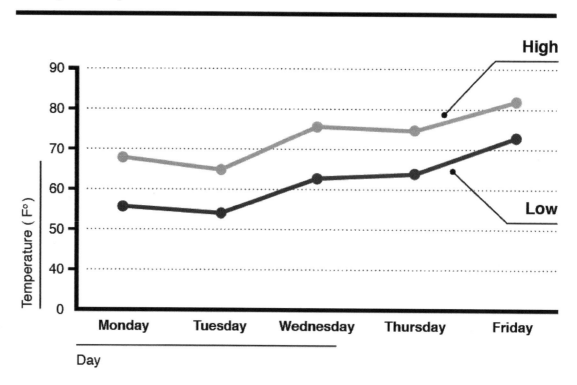

Pie charts are used to show percentages of a whole, as each category is given a piece of the pie, and together all the pieces make up a whole. They are a circular representation of data which are used to highlight numerical proportion. It is true that the arc length of each pie slice is proportional to the amount it individually represents. When a pie chart is shown, an audience can quickly make comparisons by comparing the sizes of the pieces of the pie. They can be useful for comparison between different categories. The following pie chart is a simple example of three different categories shown in comparison to each other.

Light gray represents cats, dark gray represents dogs, and the gray between those two represents other pets. As the pie is cut into three equal pieces, each value represents just more than 33 percent, or $\frac{1}{3}$ of the whole. Values 1 and 2 may be combined to represent $\frac{2}{3}$ of the whole.

In an example where the total pie represents 75,000 animals, then cats would be equal to $\frac{1}{3}$ of the total, or 25,000. Dogs would equal 25,000 and other pets would hold equal 25,000.

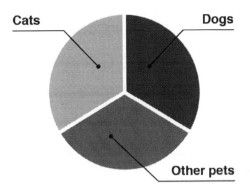

The fact that a circle is 360 degrees is used to create a pie chart. Because each piece of the pie is a percentage of a whole, that percentage is multiplied times 360 to get the number of degrees each piece represents. In the example above, each piece is 1/3 of the whole, so each piece is equivalent to 120 degrees. Together, all three pieces add up to 360 degrees.

Stacked bar graphs, also used fairly frequently, are used when comparing multiple variables at one time. They combine some elements of both pie charts and bar graphs, using the organization of bar graphs and the proportionality aspect of pie charts. The following is an example of a stacked bar graph that represents the number of students in a band playing drums, flute, trombone, and clarinet. Each bar graph is broken up further into girls and boys:

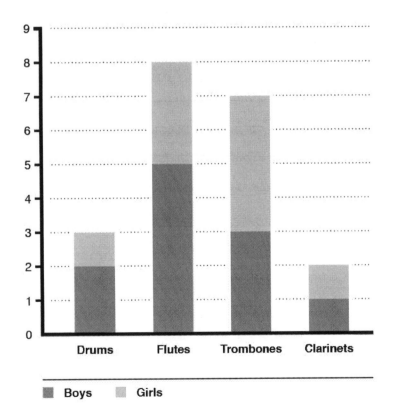

To determine how many boys play trombone, refer to the darker portion of the trombone bar, resulting in 3 students.

A **scatterplot** is another way to represent paired data. It uses Cartesian coordinates, like a line graph, meaning it has both a horizontal and vertical axis. Each data point is represented as a dot on the graph. The dots are never connected with a line. For example, the following is a scatterplot showing people's height versus age.

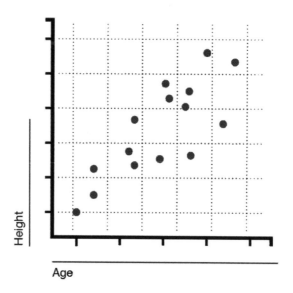

A scatterplot, also known as a **scattergram**, can be used to predict another value and to see if an association, known as a **correlation**, exists between a set of data. If the data resembles a straight line, the data is **associated.** The following is an example of a scatterplot in which the data does not seem to have an association:

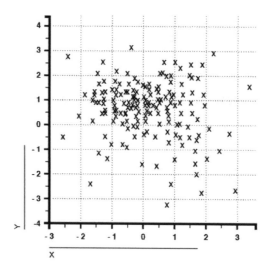

Sets of numbers and other similarly organized data can also be represented graphically. Venn diagrams are a common way to do so. A **Venn diagram** represents each set of data as a circle. The circles overlap, showing that each set of data is overlapping. A Venn diagram is also known as a **logic diagram** because it visualizes all possible logical combinations between two sets. Common elements of two sets are represented by the area of overlap.

The following is an example of a Venn diagram of two sets A and B:

Parts of the Venn Diagram

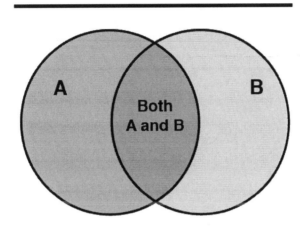

Another name for the area of overlap is the **intersection.** The intersection of A and B, $A \cap B$, contains all elements that are in both sets A and B. The **union** of A and B, $A \cup B$, contains all elements that are in either set A or set B. Finally, the **complement** of $A \cup B$ is equal to all elements that are not in either set A or set B. These elements are placed outside of the circles.

The following is an example of a Venn diagram in which 30 students were surveyed asking which type of siblings they had: brothers, sisters, or both. Ten students only had a brother, 7 students only had a sister, and 5 had both a brother and a sister. This number 5 is the intersection and is placed where the circles overlap. Two students did not have a brother or a sister. Two is therefore the complement and is placed outside of the circles.

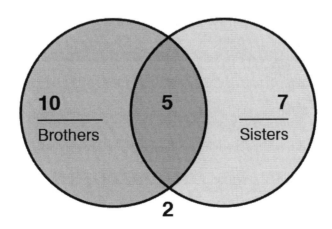

Venn diagrams can have more than two sets of data. The more circles, the more logical combinations are represented by the overlapping. The following is a Venn diagram that represents a different situation. Now, there were 30 students surveyed about the color of their socks. The innermost region represents those students that have green, pink, and blue socks on (perhaps a striped pattern). Therefore, 2 students had all three colors on their socks. In this example, all students had at least one of the three colors on their socks, so no one exists in the complement.

30 students

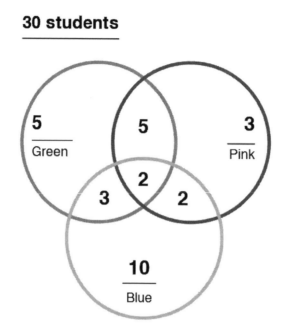

Venn diagrams are typically not drawn to scale, but if they are and their area is proportional to the amount of data it represents, it is known as an *area-proportional* Venn diagram.

Evaluating the Information in Tables, Charts, and Graphs Using Statistics

One way information can be interpreted from tables, charts, and graphs is through statistics. The three most common calculations for a set of data are the mean, median, and mode. These three are called **measures of central tendency**. Measures of central tendency are helpful in comparing two or more different sets of data. The **mean** refers to the average and is found by adding up all values and dividing the total by the number of values. In other words, the mean is equal to the sum of all values divided by the number of data entries. For example, if you bowled a total of 532 points in 4 bowling games, your mean score was $\frac{532}{4} = 133$ points per game. A common application of mean useful to students is calculating what he or she needs to receive on a final exam to receive a desired grade in a class.

The **median** is found by lining up values from least to greatest and choosing the middle value. If there's an even number of values, then the mean of the two middle amounts must be calculated to find the median. For example, the median of the set of dollar amounts $5, $6, $9, $12, and $13 is $9. The median of the set of dollar amounts $1, $5, $6, $8, $9, $10 is $7, which is the mean of $6 and $8. The **mode** is the value that occurs the most. The mode of the data set {1, 3, 1, 5, 5, 8, 10} actually refers to two numbers: 1 and 5. In this case, the data set is bimodal because it has two modes. A data set can have no mode if no amount is repeated. Another useful statistic is range. The **range** for a set of data refers to the difference between the highest and lowest value.

In some cases, some numbers in a list of data might have weights attached to them. In that case, a weighted mean can be calculated. A common application of a weighted mean is GPA. In a semester, each class is assigned a number of credit hours, its weight, and at the end of the semester each student receives a grade. To compute GPA, an A is a 4, a B is a 3, a C is a 2, a D is a 1, and an F is a 0. Consider a student that takes a 4-hour English class, a 3-hour math class, and a 4-hour history class and receives all B's. The weighted mean, GPA, is found by multiplying each grade times its weight, number of credit hours, and dividing by the total number of credit hours. Therefore, the student's GPA is:

$$\frac{3 \cdot 4 + 3 \cdot 3 + 3 \cdot 4}{11} = \frac{33}{1} = 3.0.$$

The following bar chart shows how many students attend a cycle class on each day of the week. To find the mean attendance for the week, each day's attendance can be added together,

$$10 + 7 + 6 + 9 + 8 + 14 + 4 = 58$$

and the total divided by the number of days, $58 \div 7 = 8.3$. The mean attendance for the week was 8.3 people. The median attendance can be found by putting the attendance numbers in order from least to greatest: 4, 6, 7, 8, 9, 10, 14, and choosing the middle number: 8 people. There is no mode for this set of data because no numbers repeat. The range is 10, which is found by finding the difference between the lowest number, 4, and the highest number, 14.

Cycle class attendance

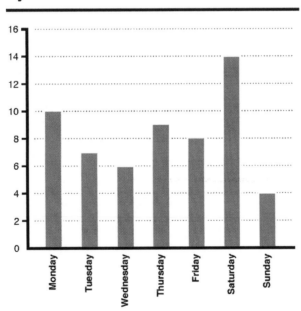

A **histogram** is a bar graph used to group data into "bins" that cover a range on the horizontal, or x-axis. Histograms consist of rectangles whose height is equal to the frequency of a specific category. The horizontal axis represents the specific categories. Because they cover a range of data, these bins have no gaps between bars, unlike the bar graph above. In a histogram showing the heights of adult golden retrievers, the bottom axis would be groups of heights, and the y-axis would be the number of dogs in each range. Evaluating this histogram would show the height of most golden retrievers as falling within a certain range. It also provides information to find the average height and range for how tall golden retrievers may grow.

The following is a histogram that represents exam grades in a given class. The horizontal axis represents ranges of the number of points scored, and the vertical axis represents the number of students. For example, approximately 33 students scored in the 60 to 70 range.

Results of the exam

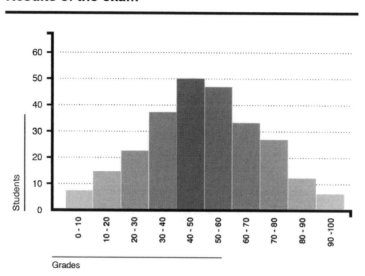

Measures of central tendency can be discussed using a histogram. If the points scored were shown with individual rectangles, the tallest rectangle would represent the mode. A **bimodal** set of data would have two peaks of equal height. Histograms can be classified as having data **skewed to the left, skewed to the right,** or **normally distributed,** which is also known as **bell-shaped.**

These three classifications can be seen in the following image:

Measures of central tendency images

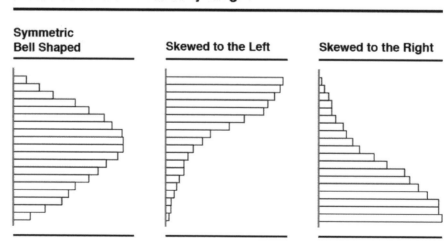

When the data follows the normal distribution, the mean, median, and mode are all very close. They all represent the most typical value in the data set. The mean is typically used as the best measure of central tendency in this case because it does include all data points. However, if the data is skewed, the mean becomes less meaningful. The median is the best measure of central tendency because it is not affected by any outliers, unlike the mean. When the data is skewed, the mean is dragged in the direction of the skew. Therefore, if the data is not normal, it is best to use the median as the measure of central tendency.

The measures of central tendency and the range may also be found by evaluating information on a line graph.

In the line graph from a previous example that showed the daily high and low temperatures, the average high temperature can be found by gathering data from each day on the triangle line. The days' highs are 82, 78, 75, 65, and 70. The average is found by adding them together to get 370, then dividing by 5 (because there are 5 temperatures). The average high for the five days is 74. If 74 degrees is found on the graph, then it falls in the middle of the values on the triangle line. The mean of the low temperature can be found in the same way.

Given a set of data, the **correlation coefficient**, r, measures the association between all the data points. If two values are correlated, there is an association between them. However, correlation does not necessarily mean causation, or that that one value causes the other. There is a common mistake made that assumes correlation implies causation. Average daily temperature and number of sunbathers are both correlated and have causation. If the temperature increases, that change in weather causes more people to want to catch some rays. However, wearing plus-size clothing and having heart disease are two variables that are correlated but do not have causation. The larger someone is, the more likely he or she is to have heart disease. However, being overweight does not cause someone to have the disease.

The value of the correlation coefficient is between -1 and 1, where -1 represents a perfect negative linear relationship, 0 represents no relationship between the two data sets, and 1 represents a perfect positive linear relationship. A negative linear relationship means that as x values increase, y values decrease. A positive linear relationship means that as x values increase, y values increase. The formula for computing the correlation coefficient is:

$$r = \frac{n \sum xy - (\sum x)(\sum y)}{\sqrt{n(\sum x^2) - (\sum x)^2} \sqrt{n(\sum y^2) - (y)^2}}$$

In this formula, n is the number of data points.

The closer r is to 1 or -1, the stronger the correlation. A correlation can be seen when plotting data. If the graph resembles a straight line, there is a correlation.

Constructing Graphs That Correctly Represent Given Data

Data is often displayed with a line graph, bar graph, or pie chart.

The line graph below shows the number of push-ups that a student did over one week:

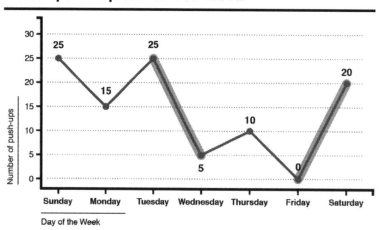

Notice that the horizontal axis displays the day of the week and the vertical axis displays the number of push-ups. A point is placed above each day of the week to show how many push-ups were done each day. For example, on Sunday the student did 25 push-ups. The line that connects the points shows how much the number of push-ups fluctuated throughout the week.

The bar graph below compares the number of people who own various types of pets:

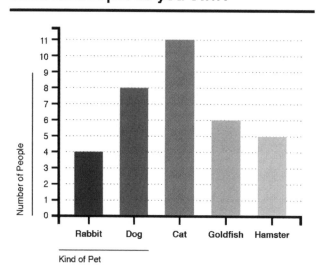

On the horizontal axis, the kind of pet is displayed. On the vertical axis, the number of people is displayed. Bars are drawn to show the number of people who own each type of pet. With the bar graph, it can quickly be determined that the fewest number of people own a rabbit and the greatest number of people own a cat.

The pie graph below displays students in a class who scored A, B, C, or D. Each slice of the pie is drawn to show the portion of the whole class that is represented by each letter grade. For example, the smallest portion represents students who scored a D. This means that the fewest number of students scored a D.

Student Grades

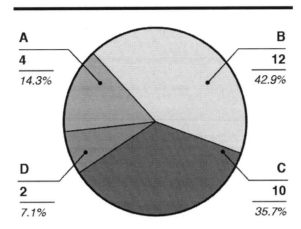

Choosing Appropriate Graphs to Display Data

Data may be displayed with a line graph, bar graph, or pie chart.

A line graph is used to display data that changes continuously over time.

A bar graph is used to compare data from different categories or groups and is helpful for recognizing relationships.

A pie chart is used when the data represents parts of a whole.

Important Features of Graphs

A **graph** is a pictorial representation of the relationship between two variables. To read and interpret a graph, it is necessary to identify important features of the graph. First, read the title to determine what data sets are being related in the graph. Next, read the axis labels and understand the scale that is used. The horizontal axis often displays categories, like years, month, or types of pets. The vertical axis often displays numerical data like amount of income, number of items sold, or number of pets owned. Check to see what increments are used on each axis. The changes on the axis may represent fives, tens, hundreds, or any increment. Be sure to note what the increment is because it will affect the interpretation of the graph. Now, locate on the graph an element of interest and move across to find the element to which it relates. For example, notice an element displayed on the horizontal axis, find that element on the graph, and then follow it across to the corresponding point on the vertical axis. Using the appropriate scale, interpret the relationship.

Explaining the Relationship between Two Variables

Independent and dependent are two types of variables that describe how they relate to each other. The **independent variable** is the variable controlled by the experimenter. It stands alone and isn't changed by other parts of the experiment. This variable is normally represented by x and is found on the horizontal, or *x*-axis, of a graph. The **dependent variable** changes in response to the independent variable. It reacts to, or depends on, the independent variable. This variable is normally represented by *y* and is found on the vertical, or *y*-axis of the graph.

The relationship between two variables, *x* and *y*, can be seen on a scatterplot.

The following scatterplot shows the relationship between weight and height. The graph shows the weight as *x* and the height as *y*. The first dot on the left represents a person who is 45 kg and approximately 150 cm tall. The other dots correspond in the same way. As the dots move to the right and weight increases, height also increases. A line could be drawn through the middle of the dots to move from bottom left to top right. This line would indicate a **positive correlation** between the variables. If the variables had a **negative correlation**, then the dots would move from the top left to the bottom right.

Height and Weight

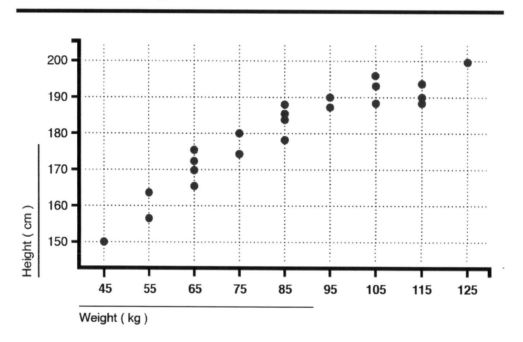

A **scatterplot** is useful in determining the relationship between two variables, but it's not required. Consider an example where a student scores a different grade on his math test for each week of the month. The independent variable would be the weeks of the month. The dependent variable would be the grades, because they change depending on the week. If the grades trended up as the weeks passed, then the relationship between grades and time would be positive. If the grades decreased as the time passed, then the relationship would be negative. (As the number of weeks went up, the grades went down.)

The relationship between two variables can further be described as strong or weak. The relationship between age and height shows a strong positive correlation because children grow taller as they grow up. In adulthood, the relationship between age and height becomes weak, and the dots will spread out. People stop growing in adulthood, and their final heights vary depending on factors like genetics and health. The closer the dots on the graph, the stronger the relationship. As they spread apart, the relationship becomes weaker. If they are too spread out to determine a correlation up or down, then the variables are said to have no correlation.

Variables are values that change, so determining the relationship between them requires an evaluation of who changes them. If the variable changes because of a result in the experiment, then it's dependent. If the variable changes before the experiment, or is changed by the person controlling the experiment, then it's the independent variable. As they interact, one is manipulated by the other. The manipulator is the independent, and the manipulated is the dependent. Once the independent and dependent variable are determined, they can be evaluated to have a positive, negative, or no correlation.

Comparing Linear Growth with Exponential Growth

Linear growth involves a quantity, the dependent variable, increasing or decreasing at a constant rate as another quantity, the independent variable, increases as well. The graph of linear growth is a straight line. Linear growth is represented as the following equation: $y = mx + b$, where m is the **slope** of the line, also known as the **rate of change**, and b is the y-intercept. If the y-intercept is 0, then the linear growth is actually known as direct variation. If the slope is positive, the dependent variable increases as the independent variable increases, and if the slope is negative, the dependent variable decreases as the independent variable increases.

Exponential growth involves a quantity, the dependent variable, changing by a common ratio every unit increase or equal interval. The equation of exponential growth is $y = a^x$ for $a > 0, a \neq 1$. The value a is known as the **base.** Consider the exponential equation $y = 2^x$. When $x = 1$, $y = 2$, and when $x = 2$, $y = 4$. For every unit increase in x, the value of the output variable doubles. Here is the graph of $y = 2^x$. Notice that as the dependent variable, y, gets very large, x increases slightly. This characteristic of this graph is why sometimes a quantity is said to be blowing up exponentially.

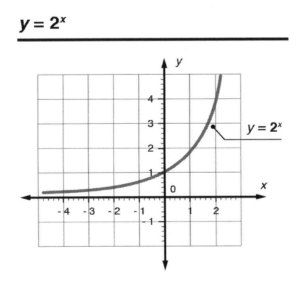

$y = 2^x$

Statistical Questions

Statistics is the branch of mathematics that deals with the collection, organization, and analysis of data. A statistical question is one that can be answered by collecting and analyzing data. When collecting data, expect variability. For example, "How many pets does Yanni own?" is not a statistical question because it can be answered in one way. "How many pets do the people in a certain neighborhood own?" is a statistical question because, to determine this answer, one would need to collect data from each person in the neighborhood, and it is reasonable to expect the answers to vary.

Identify these as statistical or not statistical:

1. How old are you?

2. What is the average age of the people in your class?

3. How tall are the students in Mrs. Jones' sixth grade class?

4. Do you like Brussels sprouts?

Questions 2 and 3 are statistical questions.

Using Statistics to Investigate Data

In statistics, measures of central tendency are measures of average. They include the mean, median, mode, and midrange of a data set. The **mean**, otherwise known as the **arithmetic average**, is found by dividing the sum of all data entries by the total number of data points. The **median** is the midpoint of the data points. If there is an odd number of data points, the median is the entry in the middle. If there is an even number of data points, the median is the mean of the two entries in the middle. The **mode** is the data point that occurs most often. Finally, the **midrange** is the mean of the lowest and highest data points. Given the spread of the data, each type of measure has pros and cons. In a **right-skewed distribution**, the bulk of the data falls to the left of the mean. In this situation, the mean is on the right of the median and the mode is on the left of the median. In a **normal distribution**, where the data are evenly distributed on both sides of the mean, the mean, median, and mode are very close to one another. In a **left-skewed distribution,** the bulk of the data falls to the right of the mean. The mean is on the left of the median and the mode is on the right of the median. Here is an example of each type of distribution:

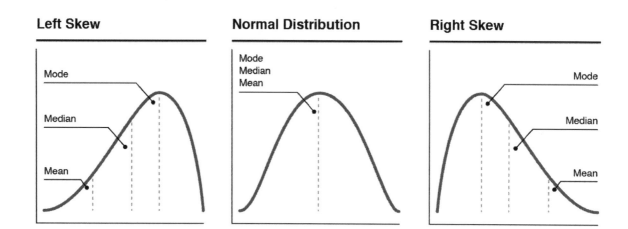

Solving Problems Involving Measures of Center and Range

A data set can be described by calculating the mean, median, and mode. These values, called **measures of center,** allow the data to be described with a single value that is representative of the data set.

The most common measure of center is the **mean**, also referred to as the **average**.

To calculate the mean,

- Add all data values together

- Divide by the sample size (the number of data points in the set)

The **median** is middle data value, so that half of the data lies below this value and half lies below the data value.

To calculate the median,

- Order the data from least to greatest

- The point in the middle of the set is the median

    o In the event that there is an even number of data points, add the two middle points and divide by 2

The **mode** is the data value that occurs most often.

To calculate the mode,

- Order the data from least to greatest

- Find the value that occurs most often

Example: Amelia is a leading scorer on the school's basketball team. The following data set represents the number of points that Amelia has scored in each game this season. Use the mean, median, and mode to describe the data.

16, 12, 26, 14, 13, 28, 14, 12, 15, 25

Solution:

Mean: $16 + 12 + 26 + 14 + 28 + 14 + 12 + 15 + 25 = 162$

$162 \div 9 = 18$

Amelia averages 18 points per game.

Median: 12, 12, 14, 14, **15**, 16, 25, 26, 28

Amelia's median score is 15.

Mode: 12, 12, 14, 14, 15, 16, 25, 26, 28

12 and 14 each occur twice in the data set, so this set has 2 modes: 12 and 14.

The **range** is the difference between the largest and smallest values in the set. In the example above, the range is $28 - 12 = 16$.

How Changes in Data Affect Measures of Center or Range

An **outlier** is a data point that lies an unusual distance from other points in the data set. Removing an outlier from a data set will change the measures of center. Removing a large outlier from a data set will decrease both the mean and the median. Removing a small outlier from a data set will increase both the mean and the median. For example, given the data set {3, 6, 8, 12, 13, 14, 60}, the data point 60 is an outlier because it is unusually far from the other points. In this data set, the mean is 16.6. Notice that this mean number is even larger than all other data points in the set except for 60. Removing the outlier, the mean changes to 9.3 and the median becomes 10. Removing an outlier will also decrease the range. In the data set above, the range is 57 when the outlier is included, but decreases to 11 when the outlier is removed.

Adding an outlier to a data set will affect the centers of measure as well. When a larger outlier is added to a data set, the mean and median increase. When a small outlier is added to a data set, the mean and median decrease. Adding an outlier to a data set will increase the range.

This does not seem to provide an appropriate measure of center when considering this data set. What will happen if that outlier is removed? Removing the extremely large data point, 60, is going to reduce the mean to 9.3. The mean decreased dramatically because 60 was much larger than any of the other data points. What would happen with an extremely low value in a data set like this one, {12, 87, 90, 95, 98, 100}? The mean of the given set is 80. When the outlier, 12, is removed, the mean should increase and should fit more closely to the other data points. Removing 12 and recalculating the mean shows that this is correct. The mean after removing 12 is 94. So, removing a large outlier will decrease the mean while removing a small outlier will increase the mean.

Data Collection Methods

Data collection can be done through surveys, experiments, observations, and interviews. A **census** is a type of survey that is done with a whole population. Because it can be difficult to collect data for an entire population, sometimes a **sample survey** is used. In this case, one would survey only a fraction of the population and make inferences about the data. Sample surveys are not as accurate as a census, but this is an easier and less expensive method of collecting data. An **experiment** is used when a researcher wants to explain how one variable causes changes in another variable. For example, if a researcher wanted to know if a particular drug affects weight loss, he would choose a treatment group that would take the drug, and another group, the control group, that would not take the drug. Special care must be taken when choosing these groups to ensure that bias is not a factor. **Bias** occurs when an outside factor influences the outcome of the research. In observational studies, the researcher does not try to influence either variable but simply observes the behavior of the subjects. Interviews are sometimes used to collect data as well. The researcher will ask questions that focus on her area of interest in order to gain insight from the participants. When gathering data through observation or interviews, it is important that the researcher is well trained so that he does not influence the results and so that the study is reliable. A study is reliable if it can be repeated under the same conditions and the same results are obtained each time.

Making Inferences About Population Parameters Based on Sample Data

In statistics, a **population** contains all subjects being studied. For example, a population could be every student at a university or all males in the United States. A **sample** consists of a group of subjects from an entire population. A sample would be 100 students at a university or 100,000 males in the United States. **Inferential statistics** is the process of using a sample to generalize information concerning populations.

Hypothesis testing is the actual process used when evaluating claims made about a population based on a sample.

A **statistic** is a measure obtained from a sample, and a **parameter** is a measure obtained from a population. For example, the mean SAT score of the 100 students at a university would be a statistic, and the mean SAT score of all university students would be a parameter.

The beginning stages of hypothesis testing starts with formulating a **hypothesis**, a statement made concerning a population parameter. The hypothesis may be true, or it may not be true. The test will answer that question. In each setting, there are two different types of hypotheses: the **null hypothesis**, written as H_0, and the **alternative hypothesis**, written as H_1. The null hypothesis represents verbally when there is not a difference between two parameters, and the alternative hypothesis represents verbally when there is a difference between two parameters.

Consider the following experiment: A researcher wants to see if a new brand of allergy medication has any effect on drowsiness of the patients who take the medication. He wants to know if the average hours spent sleeping per day increases. The mean for the population under study is 8 hours, so $\mu = 8$. In other words, the population parameter is μ, the mean. The null hypothesis is $\mu = 8$ and the alternative hypothesis is $\mu > 8$. When using a smaller sample of a population, the null hypothesis represents the situation when the mean remains unaffected and the alternative hypothesis represents the situation when the mean increases. The chosen statistical test will apply the data from the sample to actually decide whether the null hypothesis should or should not be rejected.

Dependent Versus Independent Variables

Independent variables are independent, meaning they are not changed by other variables within the context of the problem. **Dependent variables** are dependent, meaning they may change depending on how other variables change in the problem. For example, in the formula for the perimeter of a fence, the length and width are the independent variables and the perimeter is the dependent variable. The formula is shown below:

$$P = 2l + 2w$$

As the width or the length changes, the perimeter may also change. The first variables to change are the length and width, which then result in a change in perimeter. The change does not come first with the perimeter and then with length and width. When comparing these two types of variables, it is helpful to ask which variable causes the change and which variable is affected by the change.

Another formula to represent this relationship is the formula for circumference show below:

$$C = \pi \times d$$

The C represents circumference and the d represents diameter. The pi symbol represents the ratio $\frac{22}{7}$ or approximately 3.14. In this formula, the diameter of the circle is the independent variable. It is the portion of the circle that changes, which changes the circumference as a result. The circumference is the variable that is being changed by the diameter, so it is called the dependent variable. It depends on the value of the diameter.

Another place to recognize independent and dependent variables can be in experiments. A common experiment is one where the growth of a plant is tested based on the amount of sunlight it receives. Each plant in the experiment is given a different amount of sunlight, but the same amount of other nutrients like light and water. The growth of the plants is measured over a given time period and the results show how much sunlight is best for plants. In this experiment, the independent variable is the amount of sunlight that each plant receives. The dependent variable is the growth of each plant. The growth depends on the amount of sunlight, which gives reason for the distinction between independent and dependent variables.

Interpreting Probabilities Relative to Likelihood of Occurrence

Probability describes how likely it is that an event will occur. Probabilities are always a number from 0 to 1. If an event has a high likelihood of occurrence, it will have a probability close to 1. If there is only a small chance that an event will occur, the likelihood is close to 0. A fair six-sided die has one of the numbers 1, 2, 3, 4, 5, and 6 on each side. When this die is rolled there is a one in six chance that it will land on 2. This is because there are six possibilities and only one side has a 2 on it. The probability then is $\frac{1}{6}$ or .167. The probability of rolling an even number from this die is three in six, or ½ or .5. This is because there are three sides on the die with even numbers (2, 4, 6), and there are six possible sides. The probability of rolling a number less than 10 is one because every side of the die has a number less than 6, so this is certain to occur. On the other hand, the probability of rolling a number larger than 20 is zero. There are no numbers greater than 20 on the die, so it is certain that this will not occur, thus the probability is zero.

If a teacher says that the probability of anyone passing her final exam is .2, is it highly likely that anyone will pass? No, the probability of anyone passing her exam is low because .2 is closer to 0 than to 1. If another teacher is proud that the probability of students passing his class is .95, how likely is it that a student will pass? It is highly likely that a student will pass because the probability, .95, is very close to 1.

Geometry and Measurement

Lines, Rays, Line Segments, Parallel Lines, and Perpendicular Lines

Geometric figures can be identified by matching the definition with the object. For example, a **line segment** is made up of two endpoints and the line drawn between them. A **ray** is made up of one endpoint and one extending side that goes on forever. A line has no endpoints and two sides that extend on forever. These three geometric entities are shown below. What happens at A and B determines the name of each figure.

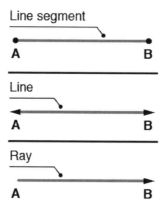

Parallel and perpendicular lines are made up of two lines, like the second figure above. They are distinguished from each other by how the two lines interact. **Parallel** lines run alongside one another, but they never intersect; the distance between them always remains the same. **Perpendicular** lines intersect at a 90-degree, or a right, angle. An example of these two sets of lines is show below. Also shown in the figure are nonexamples of these two types of lines. Because the first set of lines, in the top left corner, will eventually intersect if they continue, they are not parallel. In the second set, the lines run in the same direction and will never intersect, making them parallel. The third set, in the bottom left corner, intersect at an angle that is not right, or not 90 degrees. The fourth set is perpendicular because the lines intersect at exactly a right angle.

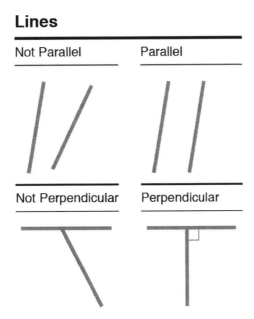

Lines

Not Parallel	Parallel
Not Perpendicular	Perpendicular

Classifying Angles Based on Their Measure

When two rays are joined together at their endpoints, an **angle** is formed. Angles can be described based on their measure. An angle whose measure is 90 degrees is described as a right angle, just as with perpendicular lines. Ninety degrees is a standard, to which other angles are compared. If an angle is less than ninety degrees, it is an **acute** angle. If it is greater than ninety degrees, it is **obtuse**. If an angle is equal to twice a right angle, or 180 degrees, it is a **straight** angle.

Examples of these types of angles are shown below:

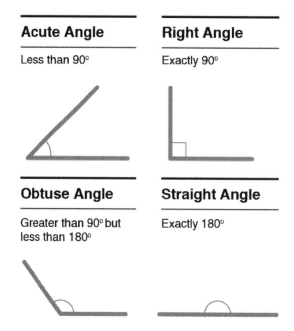

Acute Angle

Less than 90°

Right Angle

Exactly 90°

Obtuse Angle

Greater than 90° but
less than 180°

Straight Angle

Exactly 180°

A **straight angle** is equal to 180 degrees, or a straight line. If the line continues through the **vertex**, or point where the rays meet, and does not change direction, then the angle is straight. This is shown in Figure 1 below. The second figure shows an obtuse angle. Its measure is greater than ninety degrees, but less than that of a straight angle. An estimate for its measure may be 175 degrees. Figure 3 shows acute angles. The first is just less than that of a right angle. Its measure may be estimated to be 80 degrees. The second acute angle has a measure that is much smaller, at approximately 35 degrees, but it is still classified as acute because it is between zero and 90 degrees.

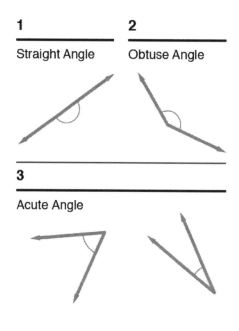

1

Straight Angle

2

Obtuse Angle

3

Acute Angle

Composing and Decomposing Two- and Three-Dimensional Shapes

Basic shapes are those polygons that are made up of straight lines and angles and can be described by their number of sides and concavity. Some examples of those shapes are rectangles, triangles, hexagons, and pentagons. These shapes have identifying characteristics on their own, but they can also be decomposed into other shapes. For example, the following can be described as one hexagon, as see in the first figure. It can also be decomposed into six equilateral triangles. The last figure shows how the hexagon can be decomposed into three rhombuses.

Decomposing a Hexagon

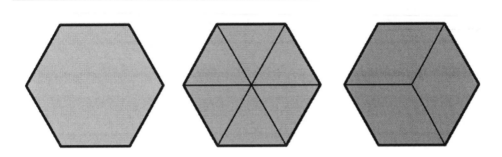

More complex shapes can be formed by combining basic shapes, or lining them up side by side. Below is an example of a house. This house is one figure all together, but can be decomposed into seven different shapes. The chimney is a parallelogram and the roof is made up of two triangles. The bottom of the house is a square alongside three triangles. There are many other ways of decomposing this house. Different shapes can be used to line up together and form one larger shape. The area for the house can be calculated by finding the individual areas for each shape, then adding them all together. For this house, there would be the area of four triangles, one square, and one parallelogram. Adding these all together would result in the area of the house as a whole. Decomposing and composing shapes is commonly done with a set of tangrams. A **tangram** is a set of shapes that includes different size triangles, rectangles, and parallelograms.

A Tangram of a House

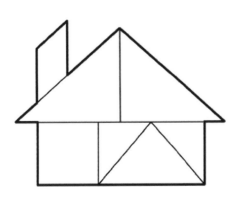

107

Shapes and Solids

Shapes are defined by their angles and number of sides. A shape with one continuous side, where all points on that side are equidistant from a center point is called a **circle**. A shape made with three straight line segments is a **triangle**. A shape with four sides is called a **quadrilateral**, but more specifically a **square, rectangle, parallelogram,** or **trapezoid,** depending on the interior angles. These shapes are two-dimensional and only made of straight lines and angles. **Solids** can be formed by combining these shapes and forming three-dimensional figures. These figures have another dimension because they add one more direction. Examples of solids may be prisms or spheres. There are four figures below that can be described based on their sides and dimensions. Figure 1 is a cone because it has three dimensions, where the bottom is a circle and the top is formed by the sides combining to one point. Figure 2 is a triangle because it has two dimensions, made up of three line segments. Figure 3 is a **cylinder** made up of two base circles and a rectangle to connect them in three dimensions. Figure 4 is an **oval** because it is one continuous line in two dimensions, not equidistant from the center.

Shapes and Solids

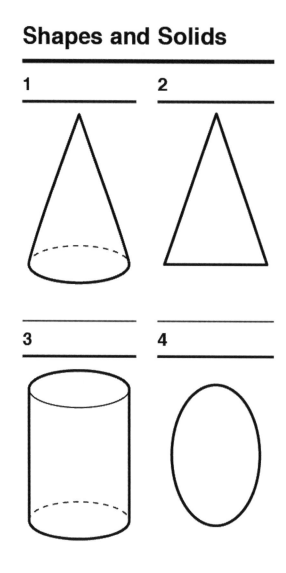

1

2

3

4

Figure 5 below is made up of squares in three dimensions, combined to make a cube. Figure 6 is a **rectangle** because it has four sides that intersect at right angles. More specifically, it can be described as a **square** because the four sides have equal measures. Figure 7 is a **pyramid** because the bottom shape is a square and the sides are all triangles. These triangles intersect at a point above the square. Figure 8 is a **circle** because it is made up of one continuous line where the points are all equidistant from one center point.

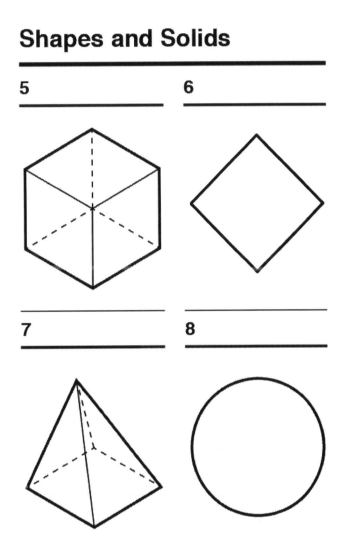

Shapes and Solids

5

6

7

8

Composition of objects is the way objects are used in conjunction with each other to form bigger, more complex shapes. For example, a rectangle and a triangle can be used together to form an arrow. Arrows can be found in many everyday scenarios, but are often not seen as the composition of two different shapes. A square is a common shape, but it can also be the composition of shapes. As seen in the second figure, there are many shapes used in the making of the one square. There are five triangles that are three different sizes. There is also one square and one parallelogram used to compose this square. These shapes can be used to compose each more complex shape because they line up, side by side, to fill in the shape with no gaps. This defines composition of shapes where smaller shapes are used to make larger, more complex ones.

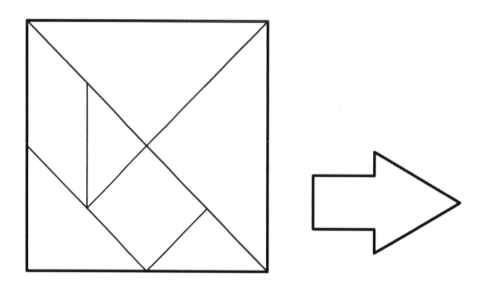

Area and Perimeter

Perimeter and area are two commonly used geometric quantities that describe objects. **Perimeter** is the distance around an object. The perimeter of an object can be found by adding the lengths of all sides. Perimeter may be used in problems dealing with lengths around objects such as fences or borders. It may also be used in finding missing lengths. If the perimeter is given, but a length is missing, subtraction is used to find the missing length. Given a square with side length s, the formula for perimeter is $P = 4s$. Given a rectangle with length l and width w, the formula for perimeter is $P = 2l + 2w$. The perimeter of a triangle is found by adding the three side lengths, and the perimeter of a trapezoid is found by adding the four side lengths. The units for perimeter are always the original units of length, such as meters, inches, miles, etc. When discussing a circle, the distance around the object is referred to as its **circumference**, not perimeter. The formula for circumference of a circle is $C = 2\pi r$, where r represents the radius of the circle. This formula can also be written as $C = \pi d$, where d represents the diameter of the circle.

Area is the two-dimensional space covered by an object. These problems may include the area of a rectangle, a yard, or a wall to be painted. Finding the area may be a simple formula, or it may require multiple formulas to be used together. The units for area are square units, such as square meters, square inches, and square miles. Given a square with side length s, the formula for its area is $A = s^2$.

Formulas for common shapes are shown below:

Shape	Formula	Graphic
Rectangle	$Area = length \times width$	
Triangle	$Area = \dfrac{1}{2} \times base \times height$	height base
Circle	$Area = \pi \times radius^2$	radius

The following formula, not as widely used as those shown above, but very important, is the area of a trapezoid:

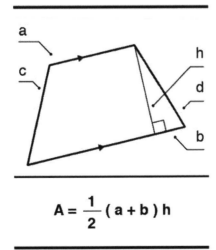

Area of a Trapezoid

$$A = \frac{1}{2}(a+b)h$$

To find the area of the shapes above, use the given dimensions of the shape in the formula. Complex shapes might require more than one formula. To find the area of the figure below, break the figure into two shapes. The rectangle has dimensions 6 cm by 7 cm. The triangle has dimensions 6 cm by 6 cm. Plug the dimensions into the rectangle formula: $A = 6 \times 7$. Multiplication yields an area of 42 cm². The triangle area can be found using the formula $A = \frac{1}{2} \times 4 \times 6$. Multiplication yields an area of 12 cm². Add the areas of the two shapes to find the total area of the figure, which is 54 cm².

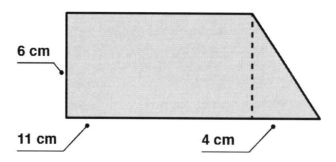

Instead of combining areas, some problems may require subtracting them, or finding the difference.

To find the area of the shaded region in the figure below, determine the area of the whole figure. Then the area of the circle can be subtracted from the whole.

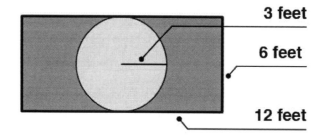

The following formula shows the area of the outside rectangle: $A = 12 \times 6 = 72\ ft^2$. The area of the inside circle can be found by the following formula: $A = \pi(3)^2 = 9\pi = 28.3\ ft^2$. As the shaded area is outside the circle, the area for the circle can be subtracted from the area of the rectangle to yield an area of 43.7 ft^2.

While some geometric figures may be given as pictures, others may be described in words. If a rectangular playing field with dimensions 95 meters long by 50 meters wide is measured for perimeter, the distance around the field must be found. The perimeter includes two lengths and two widths to measure the entire outside of the field. This quantity can be calculated using the following equation: $P = 2(95) + 2(50) = 290\ m$. The distance around the field is 290 meters.

Volume

Perimeter and area are two-dimensional descriptions; volume is three-dimensional. **Volume** describes the amount of space that an object occupies, but it's different from area because it has three dimensions instead of two. The units for volume are cubic units, such as cubic meters, cubic inches, and cubic miles. Volume can be found by using formulas for common objects such as cylinders and boxes.

The following chart shows a diagram and formula for the volume of two objects:

Shape	Formula	Diagram
Rectangular Prism (box)	$V = length \times width \times height$	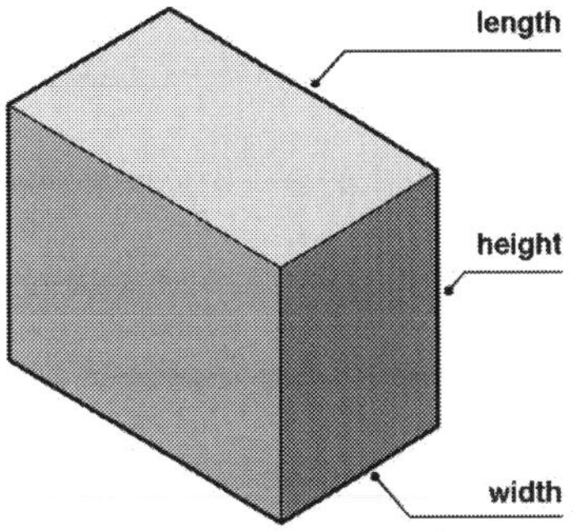
Cylinder	$V = \pi \times radius^2 \times height$	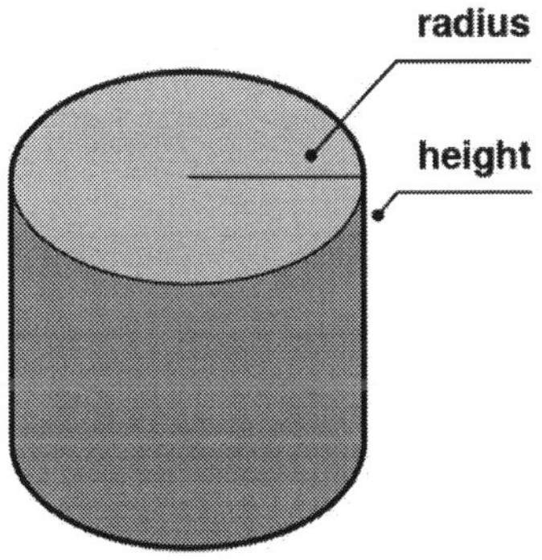

Volume formulas of these two objects are derived by finding the area of the bottom two-dimensional shape, such as the circle or rectangle, and then multiplying times the height of the three-dimensional shape. Other volume formulas include the volume of a cube with side length s: $V = s^3$; the volume of a sphere with radius r: $V = \frac{4}{3}\pi r^3$; and the volume of a cone with radius r and height h: $V = \frac{1}{3}\pi r^2 h$.

If a soda can has a height of 5 inches and a radius on the top of 1.5 inches, the volume can be found using one of the given formulas. A soda can is a cylinder. Knowing the given dimensions, the formula can be completed as follows: $V = \pi (radius)^2 \times height = \pi (1.5)^2 \times 5 = 35.325 \ inches^3$. Notice that the units for volume are inches cubed because it refers to the number of cubic inches required to fill the can.

With any geometric calculations, it's important to determine what dimensions are given and what quantities the problem is asking for. If a connection can be made between them, the answer can be found.

Other geometric quantities can include angles inside a triangle. The sum of the measures of three angles in any triangle is 180 degrees. Therefore, if only two angles are known inside a triangle, the third can be found by subtracting the sum of the two known quantities from 180. Two angles whose sum is equal to 90 degrees are known as **complementary angles**. For example, angles measuring 72 and 18 degrees are complementary, and each angle is a complement of the other. Finally, two angles whose sum is equal to 180 degrees are known as **supplementary angles**. To find the supplement of an angle, subtract the given angle from 180 degrees. For example, the supplement of an angle that is 50 degrees is 180 − 50 = 130 degrees.

These terms involving angles can be seen in many types of word problems. For example, consider the following problem: The measure of an angle is 60 degrees less than two times the measure of its complement. What is the angle's measure? To solve this, let x be the unknown angle. Therefore, its complement is $90 - x$. The problem gives that $x = 2(90 - x) - 60$. To solve for x, distribute the 2, and collect like terms. This process results in $x = 120 - 2x$. Then, use the addition property to add $2x$ to both sides to obtain $3x = 120$. Finally, use the multiplication properties of equality to divide both sides by 3 to get $x = 40$. Therefore, the angle measures 40 degrees. Also, its complement measures 50 degrees.

Right rectangular prisms are those prisms in which all sides are rectangles and all angles are right, or equal to 90 degrees. The volume for these objects can be found by multiplying the length by the width by the height. The formula is $V = lwh$. For the following prism, the volume formula is $V = 6\frac{1}{2} \times 3 \times 9$. When dealing with fractional edge lengths, it is helpful to convert the length to an improper fraction. The length 6 ½ cm becomes $\frac{13}{2}$ cm. Then the formula becomes:

$$V = \frac{13}{2} \times 3 \times 9 = \frac{13}{2} \times \frac{3}{1} \times \frac{9}{1} = \frac{351}{2}$$

This value for volume is better understood when turned into a mixed number, which would be 175 ½ cm³.

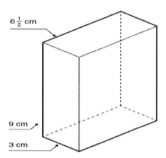

When dimensions for length are given with fractional parts, it can be helpful to turn the mixed number into an improper fraction, then multiply to find the volume, then convert back to a mixed number. When finding surface area, this conversion to improper fractions can also be helpful. The surface area can be found for the same prism above by breaking down the figure into basic shapes. These shapes are rectangles, made up of the two bases, two sides, and the front and back. The formula for the surface area uses the area for each of these shapes for the terms in the following equation:

$$SA = 6\frac{1}{2} \times 3 + 6\frac{1}{2} \times 3 + 3 \times 9 + 3 \times 9 + 6\frac{1}{2} * 9 + 6\frac{1}{2} \times 9$$

Because there are so many terms in a surface area formula and because this formula contains a fraction, it can be simplified by combining groups that are the same. Each set of numbers is used twice, to represent areas for the opposite sides of the prism. The formula can be simplified to:

$$SA = 2\left(6\frac{1}{2} \times 3\right) + 2(3 \times 9) + 2\left(6\frac{1}{2} \times 9\right)$$

$$2\left(\frac{13}{2} \times 3\right) + 2(27) + 2\left(\frac{13}{2} \times 9\right)$$

$$2\left(\frac{39}{2}\right) + 54 + 2\left(\frac{117}{2}\right)$$

$$39 + 54 + 117$$

114

210 cm²

Surface Area

Surface area is defined as the area of the surface of a figure. A **pyramid** has a surface made up of four triangles and one square. To calculate the surface area of a pyramid, the areas of each individual shape are calculated. Then the areas are added together. This method of decomposing the shape into two-dimensional figures to find area, then adding the areas, can be used to find surface area for any figure. Once these measurements are found, the area is described with square units. For example, the following figure shows a rectangular prism. The figure beside it shows the rectangular prism broken down into two-dimensional shapes, or rectangles. The area of each rectangle can be calculated by multiplying the length by the width. The area for the six rectangles can be represented by the following expression: $5 \times 6 + 5 \times 10 + 5 \times 6 + 6 \times 10 + 5 \times 10 + 6 \times 10$. The total for all these areas added together is 280m², or 280 square meters. This measurement represents the surface area because it is the area of all six surfaces of the rectangular prism.

The Net of a Rectangular Prism

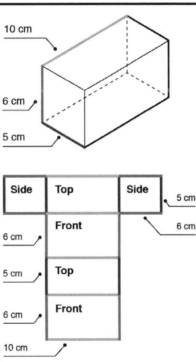

Another shape that has a surface area is a cylinder. The shapes used to make up the **cylinder** are two circles and a rectangle wrapped around between the two circles. A common example of a cylinder is a can. The two circles that make up the bases are obvious shapes. The rectangle can be more difficult to see, but the label on a can will help illustrate it. When the label is removed from a can and laid flat, the shape is a rectangle. When the areas for each shape are needed, there will be two formulas. The first is the area for the circles on the bases. This area is given by the formula $A = \pi r^2$. There will be two of these areas. Then the area of the rectangle must be determined. The width of the rectangle is equal to the height of the can, h. The length of the rectangle is equal to the circumference of the base circle, $2\pi r$. The area for the rectangle can be found by using the formula $A = 2\pi r \times h$. By adding the two areas for the bases and the area of the rectangle, the surface area of the cylinder can be found, described in units squared.

115

Circles

The formula for area of a circle is $A = \pi r^2$ and therefore, formula for area of a sector is $\pi r^2 \frac{A}{360}$, a fraction of the entire area of the circle. If the radius of a circle and arc length is known, the central angle measurement in degrees can be found by using the formula $\frac{360 \cdot arclength}{2\pi r}$. If the desired central angle measurement is in radians, the formula for the central angle measurement is much simpler as $\frac{arc\ length}{r}$.

A **chord** of a circle is a straight-line segment that connects any two points on a circle. The line segment does not have to travel through the center, as the diameter does. Also, note that the chord stops at the circumference of the circle. If it did not stop and extended toward infinity, it would be known as a **secant line**. The following shows a diagram of a circle with a chord shown by the dotted line. The radius is r and the central angle is A.

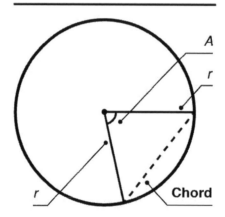

A Circle with a Chord

Chord Length $= 2\ r \sin\frac{A}{2}$

One formula for chord length can be seen in the diagram, and is equal to $2r\sin\frac{A}{2}$, where A is the central angle. Another formula for chord length is: chord length $= 2\sqrt{r^2 - D^2}$, where D is equal to the distance from the chord to the center of the circle. This formula is basically a version of the Pythagorean Theorem.

Formulas for chord lengths vary based on what type of information is known. If the radius and central angle are known, the first formula listed above should be used by plugging the radius and angle in directly. If the radius and the distance from the center to the chord are known, the second formula listed previously should be used.

Many theorems exist between arc lengths, angle measures, chord lengths, and areas of sectors. For instance, when two chords intersect in a circle, the product of the lengths of the individual line segments are equal. For instance, in the following diagram, $A \times B = C \times D$:

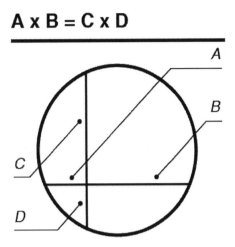

The Pythagorean Theorem

Within right triangles, trigonometric ratios can be defined for the acute angle within the triangle. Consider the following right triangle. The side across from the right angle is known as the **hypotenuse**, the acute angle being discussed is labeled θ, the side across from the acute angle is known as the **opposite** side, and the other side is known as the **adjacent** side.

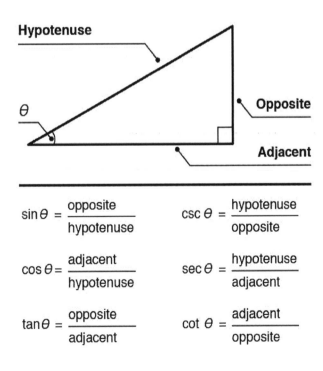

$$\sin\theta = \frac{\text{opposite}}{\text{hypotenuse}} \qquad \csc\theta = \frac{\text{hypotenuse}}{\text{opposite}}$$

$$\cos\theta = \frac{\text{adjacent}}{\text{hypotenuse}} \qquad \sec\theta = \frac{\text{hypotenuse}}{\text{adjacent}}$$

$$\tan\theta = \frac{\text{opposite}}{\text{adjacent}} \qquad \cot\theta = \frac{\text{adjacent}}{\text{opposite}}$$

The six trigonometric ratios are shown above as well. "Sin" is short for sine, "cos" is short for cosine, "tan" is short for tangent, "csc" is short for cosecant, "sec" is short for secant, and "cot" is short for cotangent. A mnemonic device exists that is helpful to remember the ratios. SOHCAHTOH stands for Sine = Opposite/Hypotenuse, Cosine = Adjacent/Hypotenuse, and Tangent = Opposite/Adjacent. The other three trigonometric ratios are reciprocals of sine, cosine, and tangent because $\csc\theta = \frac{1}{\sin\theta}$, $\sec\theta = \frac{1}{\cos\theta}$, and $\cot\theta = \frac{1}{\tan\theta}$.

The **Pythagorean Theorem** is an important relationship between the three sides of a right triangle. It states that the square of hypotenuse is equal to the sum of the squares of the other two sides. When using the Pythagorean Theorem, the hypotenuse is labeled as side c, the opposite is labeled as side a, and the adjacent side is side b. The theorem can be seen in the following diagram:

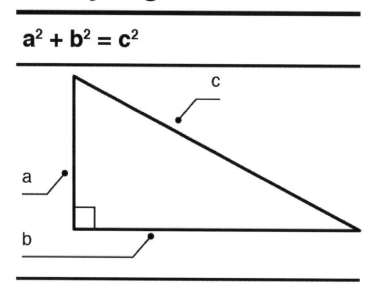

Both the trigonometric ratios and Pythagorean Theorem can be used in problems that involve finding either a missing side or missing angle of a right triangle. Look to see what sides and angles are given and select the correct relationship that will assist in finding the missing value. These relationships can also be used to solve application problems involving right triangles. Often, it is helpful to draw a figure to represent the problem to see what is missing.

Solving for Missing Values in Triangles, Circles, and Other Figures

Solving for missing values in shapes requires knowledge of the shape and its characteristics. For example, a triangle has three sides and three angles that add up to 180 degrees. If two angle measurements are given, the third can be calculated. For the triangle below, the one given angle has a measure of 55 degrees. The missing angle is *x*. The third angle is labeled with a square, which indicates a measure of 90 degrees. Because all angles must sum to 180 degrees, the following equation can be used to find the missing *x*-value: $55 + 90 + x = 180$. Adding the two given angles and subtracting the total from 180, the missing angle is found to be 35 degrees.

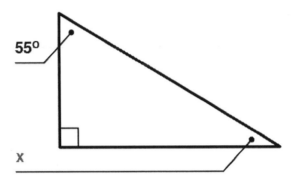

A similar problem can be solved with circles. If the radius is given, but the circumference is unknown, it can be calculated based on the formula $C = 2\pi r$. This example can be used in the figure below. The radius can be substituted for *r* in the formula. Then the circumference can be found as $C = 2\pi \times 8 = 16\pi = 50.24 \, cm$.

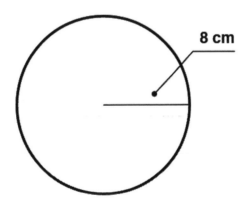

Other figures that may have missing values could be the length of a square, given the area, or the perimeter of a rectangle, given the length and width. All of the missing values can be found by first identifying all the characteristics that are known about the shape, then looking for ways to connect the missing value to the given information.

Congruency and Similarity

Two figures are **congruent** if they have the same shape and same size, meaning same angle measurements and equal side lengths. Two figures are **similar** if they have the same angle measurement but not side lengths. Basically, angles are congruent in similar triangles and their side lengths are constant multiplies of each other. Proving two shapes are similar involves showing that all angles are the same; proving two shapes are congruent involves showing that all angles are the same and that all sides are the same. If two pairs of angles are congruent in two triangles, then those triangles are similar because their third angle has to be equal due to the fact that all three angles add up to 180 degrees.

There are five main theorems that are used to show triangles are congruent. Each theorem involves showing different combinations of sides and angles are the same in two triangles, which proves the triangles are congruent. The **side-side-side (SSS)** theorem states that if all sides are equal in two triangles, the triangles are congruent. The **side-angle-side (SAS)** theorem states that if two pairs of sides are equal and the included angles are congruent in two triangles, then the triangles are congruent. Similarly, the **angle-side-angle (ASA)** theorem states that if two pairs of angles are congruent and the included side lengths are equal in two triangles, the triangles are similar. The **angle-angle-side (AAS)** theorem states that two triangles are congruent if they have two pairs of congruent angles and a pair of corresponding equal side lengths that are not included. Finally, the **hypotenuse-leg (HL)** theorem states that if two right triangles have equal hypotenuses and an equal pair of shorter sides, the triangles are congruent. An important item to note is that **angle-angle-angle (AAA)** is not enough information to have congruence because if three angles are equal in two triangles, the triangles can only be described as similar.

Similarity, Right Triangles, and Trigonometric Ratios

Within two similar triangles, corresponding side lengths are proportional, and angles are equal. In other words, regarding corresponding sides in two similar triangles, the ratio of side lengths is the same. Recall that the SAS theorem for similarity states that if an angle in one triangle is congruent to an angle in a second triangle, and the lengths of the sides in both triangles are proportional, then the triangles are similar. Also, because the ratio of two sides in two similar right triangles is the same, the trigonometric ratios in similar right triangles are always going to be equal.

If two triangles are similar, and one is a right triangle, the other is a right triangle. The definition of similarity ensures that each triangle has a 90-degree angle. In a similar sense, if two triangles are right triangles containing a pair of equal acute angles, the triangles are similar because the third pair of angles must be equal as well. However, right triangles are not necessarily always similar.

The following triangles are similar:

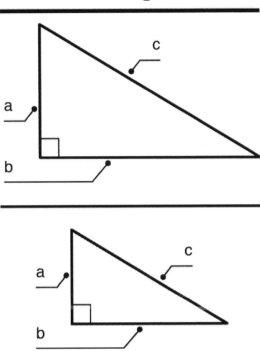

Similar Triangles

It is not apparent at first glance; however, theorems can be used to show similarity. The Pythagorean Theorem can be used to find the missing side lengths in both triangles. In the larger triangle, the missing side is the hypotenuse, c. Therefore, $9^2 + 12^2 = c^2$. This equation is equivalent to $225 = c^2$, so taking the square root of both sides results in the positive root $c = 15$. In the other triangle, the Pythagorean Theorem can be used to find the missing side length b. The theorem shows that $6^2 + b^2 = 10^2$, and b is then solved for to obtain $b = 8$. The ratio of the sides in the larger triangle to the sides in the smaller triangle is the same value, 1.5. Therefore, the sides are proportional. Because they are both right triangles, they have a congruent angle. The SAS theorem for similarity can be used to show that these two triangles are similar.

Sine and Cosine of Complementary Angles

Two **complementary angles** add up to 90 degrees, or π radians. Within a right triangle, the sine of an angle is equal to the ratio of the side opposite the angle to the hypotenuse and the cosine of an angle is equal to the ratio of the side adjacent to the angle to the hypotenuse. Within a right triangle, there is a right angle and because the sum of all angles in a triangle is 180 degrees, the two other angles add up to 90 degrees, and are therefore complementary. Consider the following right triangle with angles A, B, and C, and sides a, b, and c.

Consider the following right triangle with angles A, B, and C, and sides a, b, and c.

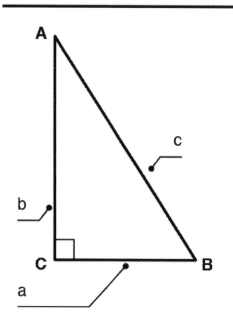

It is true by using such ratios described above that $\sin B = \frac{b}{c}$ and $\cos B = \frac{a}{c}$. Also, it is true that $\cos A = \frac{b}{c}$ and $\sin A = \frac{a}{c}$. Therefore, $\sin B = \cos A$ and $\cos B = \sin A$. A and B are complementary angles, so given two complementary angles, the sine of one equals the cosine of the other, and the cosine of one equals the sine of the other. Given the two complementary angles 30 degrees and 60 degrees, $\sin 30 = \frac{1}{2}$, $\cos 60 = \frac{1}{2}$, $\cos 30 = \frac{\sqrt{3}}{2}$, and $\sin 60 = \frac{\sqrt{3}}{2}$. Either a calculator set in degrees mode, a unit circle, or the Pythagorean Theorem could be used to find all of these values.

Representing Three-Dimensional Figures with Nets

The **net of a figure** is the resulting two-dimensional shapes when a three-dimensional shape is broken down. For example, the net of a cone is shown below. The base of the cone is a circle, shown at the bottom. The rest of the cone is a quarter of a circle. The bottom is the circumference of the circle, while the top comes to a point. If the cone is cut down the side and laid out flat, these would be the resulting shapes.

The Net of a Cone

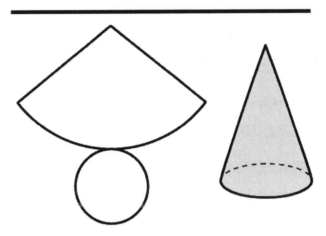

A net for a pyramid is shown in the figure below. The base of the pyramid is a square. The four shapes coming off the pyramid are triangles. When built up together, folding the triangles to the top results in a pyramid.

The Net of a Pyramid

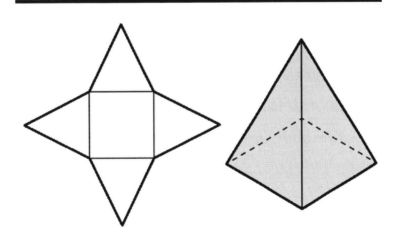

One other net for a figure is the one shown below for a cylinder. When the cylinder is broken down, the bases are circles and the side is a rectangle wrapped around the circles. The circumference of the circle turns into the length of the rectangle.

The circumference of the circle is equal to the length of the rectangle

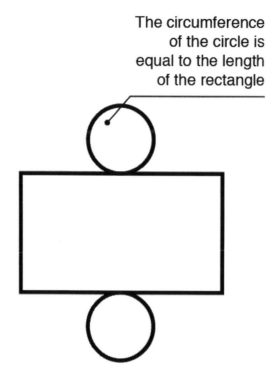

Nets can be used in calculating different values for given shapes. One useful way to calculate surface area is to find the net of the object, then find the area of each of the shapes and add them together. Nets are also useful in composing shapes and decomposing objects so as to view how objects connect and can be used in conjunction with each other.

Using Nets to Determine the Surface Area of Three-Dimensional Figures

Surface area of three-dimensional figures is the total area of each of the faces of the figures. Nets are used to lay out each face of an object. The following figure shows a triangular prism. The bases are triangles and the sides are rectangles. The second figure shows the net for this triangular prism. The dimensions are labeled for each of the faces of the prism. The area for each of the two triangles can be determined by the formula $A = \frac{1}{2}bh = \frac{1}{2} \times 8 \times 9 = 36cm^2$. The rectangle areas can be described by the equation $A = lw = 8 \times 5 + 9 \times 5 + 10 \times 5 = 40 + 45 + 50 = 135cm^2$. The area for the triangles can be multiplied by two, then added to the rectangle areas to yield a total surface are of $207cm^2$.

A Triangular Prism and Its Net

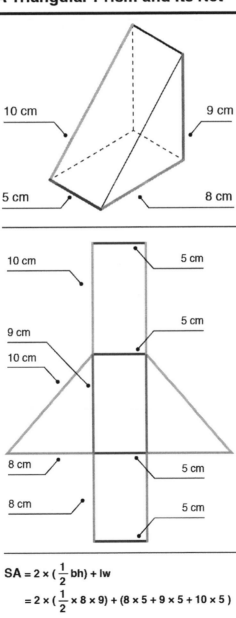

$SA = 2 \times (\frac{1}{2} bh) + lw$

$\qquad = 2 \times (\frac{1}{2} \times 8 \times 9) + (8 \times 5 + 9 \times 5 + 10 \times 5)$

$\qquad = 207cm^2$

Determining How Changes to Dimensions Change Area and Volume

When the dimensions of an object change, the area and volume are subject to change also. For example, the following rectangle has an area of 98 square centimeters. If the length is increased by 2, to be 16 cm, then the area becomes $A = 7 \times 16 = 112cm^2$. The area increased by 14 cm, or twice the width because there were two more widths of 7 cm.

For the volume of an object, there are three dimensions. The given prism has a volume of $V = 4 \times 12 \times 3 = 144m^2$. If the height is increased by 3, the volume becomes $V = 4 \times 12 \times 6 = 288m^2$. The increase of 3 for the height, or doubling of the height, resulted in a volume that was doubled. From the original, if the width was doubled, the volume would be $V = 8 \times 12 \times 3 = 288m^2$. When the width doubled, the volume was doubled also. The same increase in volume would result if the length was doubled.

Solving Problems by Plotting Points and Drawing Polygons in the Coordinate Plane

Shapes can be plotted on the coordinate plane to identify location of each vertex and the length of each side. The original shape is seen in the figure below in the first quadrant. The length is 6 and the width is 4. The reflection of this rectangle is in the fourth quadrant. A reflection across the y-axis can be found by determining each point's distance to the y-axis and moving it that same distance on the opposite side. For example, the point C is located at (2, 1). The reflection of this point moves to (–2, 1) when reflected across the y-axis. The original point A is located at (2, 5), and the reflection across the y-axis is located at (–2, 5). It is evident that the reflection across the y-axis changes the sign of the x-coordinate. A reflection across the x-axis changes the sign on the y-coordinate, as seen in the reflected figure below. Other translations can be found using the coordinate plane, such as rotations around the origin and reflections across the y-axis.

Solving Problems Involving Elapsed Time, Money, Length, Volume, and Mass

To solve problems, follow these steps: Identify the variables that are known, decide which equation should be used, substitute the numbers, and solve. To solve an equation for the amount of time that has elapsed since an event, use the equation T = L – E where T represents the elapsed time, L represents the later time, and E represents the earlier time. For example, the Minnesota Vikings have not appeared in the Super Bowl since 1976. If the year is now 2018, how long has it been since the Vikings were in the Super Bowl? The later time, L, is 2018, E = 1976 and the unknown is T. Substituting these numbers, the equation is T = 2018 – 1976, and so T = 42. It has been 42 years since the Vikings have appeared in the Super Bowl.

Questions involving total cost can be solved using the formula, $C = I + T$ where C represents the total cost, I represents the cost of the item purchased, and T represents the tax amount.

To find the length of a rectangle given the area = 32 square inches and width = 8 inches, the formula $A = L \times W$ can be used. Substitute 32 for A and substitute 8 for w, giving the equation $32 = L \times 8$. This equation is solved by dividing both sides by 8 to find that the length of the rectangle is 4. The formula for volume of a rectangular prism is given by the equation $V = L \times W \times H$. If the length of a rectangular juice box is 4 centimeters, the width is 2 centimeters, and the height is 8 centimeters, what is the volume of this box? Substituting in the formula we find $V = 4 \times 2 \times 8$, so the volume is 64 cubic centimeters. In a similar fashion as those previously shown, the mass of an object can be calculated given the formula, $Mass = Density \times Volume$.

Measuring and Comparing Lengths of Objects

Lengths of objects can be measured using tools such as rulers, yard sticks, meter sticks, and tape measures. Typically, a ruler measures 12 inches, or one foot. For this reason, a ruler is normally used to measure lengths smaller than or just slightly more than 12 inches. Rulers may represent centimeters instead of inches. Some rulers have inches on one side and centimeters on the other. Be sure to recognize what units you are measuring in. The standard ruler measurements are divided into units of 1 inch and normally broken down to $\frac{1}{2}, \frac{1}{4}, \frac{1}{8}$, and even $\frac{1}{16}$ of an inch for more precise measurements. If measuring in centimeters, the centimeter is likely divided into tenths. To measure the size of a picture, for purposes of buying a frame, a ruler is helpful. If the picture is very large, a yardstick, which measures 3 feet and normally is divided into feet and inches, might be useful. Using metric units, the meter stick measures 1 meter and is divided into 100 centimeters. To measure the size of a window in a home, either a yardstick or meter stick would work. To measure the size of a room, though, a tape measure would be the easiest tool to use. Tape measures can measure up to 10 feet, 25 feet, or more.

Comparing Relative Sizes of U.S. Customary Units and Metric Units

Measuring length in United States customary units is typically done using inches, feet, yards, and miles. When converting among these units, remember that 12 inches = 1 foot, 3 feet = 1 yard, and 5,280 feet = 1 mile. Common customary units of weight are ounces and pounds. The conversion needed is 16 ounces = 1 pound. For customary units of volume ounces, cups, pints, quarts, and gallons are typically used. For conversions, use 8 ounces = 1 cup, 2 cups = 1 pint, 2 pints = 1 quart, and 4 quarts = 1 gallon. For measuring lengths in metric units, know that 100 centimeters = 1 meter, and 1,000 meters = 1 kilometer. For metric units of measuring weights, grams and kilograms are often used. Know that 1,000 grams = 1 kilogram when making conversions. For metric measures of volume, the most common units are milliliters and liters. Remember that 1,000 milliliters = 1 liter.

Converting Within and Between Standard and Metric Systems

When working with dimensions, sometimes the given units don't match the formula, and conversions must be made. The metric system has base units of meter for length, kilogram for mass, and liter for liquid volume. This system expands to three places above the base unit and three places below. These places correspond with prefixes with a base of 10.

The following table shows the conversions:

kilo-	hecto-	deka-	base	deci-	centi-	milli-
1,000 times the base	100 times the base	10 times the base		1/10 times the base	1/100 times the base	1/1000 times the base

To convert between units within the metric system, values with a base ten can be multiplied. The decimal can also be moved in the direction of the new unit by the same number of zeros on the number. For example, 3 meters is equivalent to .003 kilometers. The decimal moved three places (the same number of zeros for kilo-) to the left (the same direction from base to kilo-). Three meters is also equivalent to 3,000 millimeters. The decimal is moved three places to the right because the prefix milli- is three places to the right of the base unit.

The English Standard system used in the United States has a base unit of foot for length, pound for weight, and gallon for liquid volume. These conversions aren't as easy as the metric system because they aren't a base ten model. The following table shows the conversions within this system:

Length	Weight	Capacity
1 foot (ft) = 12 inches (in) 1 yard (yd) = 3 feet 1 mile (mi) = 5,280 feet 1 mile = 1,760 yards	1 pound (lb) = 16 ounces (oz) 1 ton = 2,000 pounds	1 tablespoon (tbsp) = 3 teaspoons (tsp) 1 cup (c) = 16 tablespoons 1 cup = 8 fluid ounces (oz) 1 pint (pt) = 2 cups 1 quart (qt) = 2 pints 1 gallon (gal) = 4 quarts

When converting within the English Standard system, most calculations include a conversion to the base unit and then another to the desired unit. For example, take the following problem: 3 $quarts$ = ___ $cups$. There is no straight conversion from quarts to cups, so the first conversion is from quarts to pints. There are 2 pints in 1 quart, so there are 6 pints in 3 quarts. This conversion can be solved as a proportion: $\frac{3\ qt}{x} = \frac{1\ qt}{2\ pints}$. It can also be observed as a ratio 2:1, expanded to 6:3. Then the 6 pints must be converted to cups. The ratio of pints to cups is 1:2, so the expanded ratio is 6:12. For 6 pints, the measurement is 12 cups. This problem can also be set up as one set of fractions to cancel out units. It begins with the given information and cancels out matching units on top and bottom to yield the answer. Consider the following expression:

$$\frac{3\ quarts}{1} \times \frac{2\ pints}{1\ quart} \times \frac{2\ cups}{1\ pint}$$

It's set up so that units on the top and bottom cancel each other out:

$$\frac{3\ \cancel{quarts}}{1} \times \frac{2\ \cancel{pints}}{1\ \cancel{quart}} \times \frac{2\ cups}{1\ \cancel{pint}}$$

The numbers can be calculated as $3 \times 3 \times 2$ on the top and 1 on the bottom. It still yields an answer of 12 cups.

This process of setting up fractions and canceling out matching units can be used to convert between standard and metric systems. A few common equivalent conversions are 2.54 cm = 1 inch, 3.28 feet = 1 meter, and 2.205 pounds = 1 kilogram. Writing these as fractions allows them to be used in conversions. For the fill-in-the-blank problem 5 meters = ____ feet, an expression using conversions starts with the expression $\frac{5\ meters}{1} \times \frac{3.28\ feet}{1\ meter}$, where the units of meters will cancel each other out, and the final unit is feet. Calculating the numbers yields 16.4 feet. This problem only required two fractions. Others may require longer expressions, but the underlying rule stays the same. When there's a unit on the top of the fraction that's the same as the unit on the bottom, then they cancel each other out. Using this logic and the conversions given above, many units can be converted between and within the different systems.

The conversion between Fahrenheit and Celsius is found in a formula:

$$°C = (°F - 32) \times \frac{5}{9}$$

For example, to convert 78 °F to Celsius, the given temperature would be entered into the formula: °C = $(78 - 32) \times \frac{5}{9}$. Solving the equation, the temperature comes out to be 25.56 °C. To convert in the other direction, the formula becomes: °F = °C $\times \frac{9}{5} + 32$. Remember the order of operations when calculating these conversions.

Practice Questions

1. What is the solution to the equation $3(x + 2) = 14x - 5$?
   a. $x = 1$
   b. $x = -1$
   c. $x = 0$
   d. All real numbers
   e. No solution

2. What is the solution to the equation $10 - 5x + 2 = 7x + 12 - 12x$?
   a. $x = 12$
   b. $x = 1$
   c. $x = 0$
   d. All real numbers
   e. No solution

3. Which of the following is the result when solving the equation $4(x + 5) + 6 = 2(2x + 3)$?
   a. $x = 6$
   b. $x = 1$
   c. $x = 26$
   d. All real numbers
   e. No solution

4. How many cases of cola can Lexi purchase if each case is $3.50 and she has $40?
   a. 10
   b. 12
   c. 11.4
   d. 11
   e. 12.5

5. Two consecutive integers exist such that the sum of three times the first and two less than the second is equal to 411. What are those integers?
   a. 103 and 104
   b. 104 and 105
   c. 102 and 103
   d. 100 and 101
   e. 101 and 102

6. Erin and Katie work at the same ice cream shop. Together, they always work less than 21 hours a week. In a week, if Katie worked two times as many hours as Erin, how many hours could Erin work?
   a. Less than 7 hours
   b. Less than or equal to 7 hours
   c. More than 7 hours
   d. Less than 8 hours
   e. More than 8 hours

7. From the chart below, which two types of movies are preferred by more men than women?

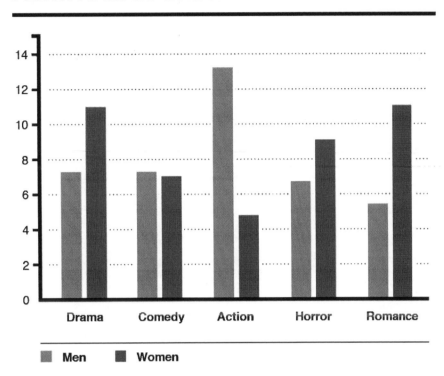

Preferred Movie Genres

a. Comedy and Action
b. Drama and Comedy
c. Action and Horror
d. Action and Romance
e. Romance and Comedy

8. Which type of graph best represents a continuous change over a period of time?
   a. Stacked bar graph
   b. Bar graph
   c. Pie graph
   d. Histogram
   e. Line graph

9. Using the graph below, what is the mean number of visitors for the first 4 hours?

Museum Visitors

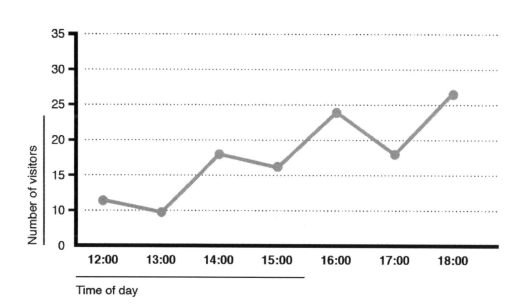

a. 12
b. 13
c. 14
d. 15
e. 16

10. What is the mode for the grades shown in the chart below?

Science Grades	
Jerry	65
Bill	95
Anna	80
Beth	95
Sara	85
Ben	72
Jordan	98

a. 65
b. 33
c. 95
d. 90
e. 84.3

11. What type of relationship is there between age and attention span as represented in the graph below?

Attention Span

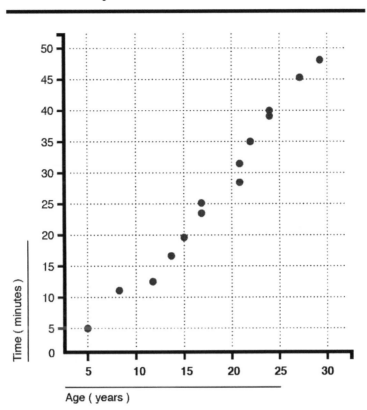

a. No correlation
b. Positive correlation
c. Negative correlation
d. Weak correlation
e. Inverse correlation

12. What is the area of the shaded region?

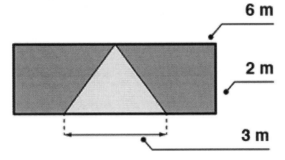

a. 9 m²
b. 12 m²
c. 6 m²
d. 8 m²
e. 4.5 m²

13. What is the volume of the cylinder below?

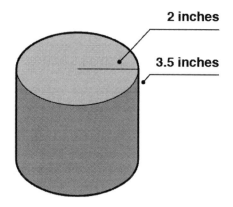

2 inches

3.5 inches

    a. 18.84 in³
    b. 45.00 in³
    c. 70.43 in³
    d. 43.96 in³
    e. 21.98 in³

14. How many kiloliters are in 6 liters?
    a. 6,000
    b. 600
    c. 0.006
    d. 0.0006
    e. 0.06

15. How many centimeters are in 3 feet? (Note: 2.54cm = 1 inch)
    a. 0.635
    b. 1.1811
    c. 14.17
    d. 7.62
    e. 91.44

16. Which expression is equivalent to $\sqrt[4]{x^6} - \frac{x}{x^3} + x - 2$?
    a. $x^{\frac{3}{2}} - x^2 + x - 2$
    b. $x^{\frac{2}{3}} - x^{-2} + x - 2$
    c. $x^{\frac{3}{2}} - \frac{1}{x^2} + x - 2$
    d. $x^{\frac{2}{3}} - \frac{1}{x^2} + x - 2$
    e. $x^{\frac{1}{3}} - \frac{1}{x^2} + x - 2$

17. How many possible positive zeros does the polynomial function $f(x) = x^4 - 3x^3 + 2x + x - 3$ have?
    a. 4
    b. 5
    c. 2
    d. 1
    e. 3

18. Which of the following is equivalent to $16^{\frac{1}{4}}16^{\frac{1}{2}}$?
   a. 8
   b. 16
   c. 4
   d. 4,096
   e. 64

19. What is the solution to the following linear inequality?
$$7 - \frac{4}{5}x < \frac{3}{5}$$
   a. $(-\infty, 8)$
   b. $(8, \infty)$
   c. $[8, \infty)$
   d. $(-\infty, 8]$
   e. $(-\infty, \infty)$

20. Triple the difference of five and a number is equal to the sum of that number and 5. What is the number?
   a. 5
   b. 2
   c. 5.5
   d. 2.5
   e. 1

21. Which of the following is perpendicular to the line $4x + 7y = 23$?
   a. $y = -\frac{4}{7}x + 23$
   b. $y = \frac{7}{4}x - 12$
   c. $4x + 7y = 14$
   d. $y = -\frac{7}{4}x + 11$
   e. $y = \frac{4}{7}x - 12$

22. The following set represents the test scores from a university class: {35, 79, 80, 87, 87, 90, 92, 95, 95, 98, 99}. If the outlier is removed from this set, which of the following is TRUE?
   a. The mean and the median will decrease.
   b. The mean and the median will increase.
   c. The mean and the mode will increase.
   d. The mean and the mode will decrease.
   e. The mean, median, and mode will increase.

23. The mass of the moon is about 7.348×10^{22} kilograms and the mass of Earth is 5.972×10^{24} kilograms. How many times GREATER is Earth's mass than the moon's mass?
   a. 8.127×10^1
   b. 8.127
   c. 812.7
   d. 8.127×10^{-1}
   e. 0.8127

24. What is the equation of the line that passes through the two points (-3, 7) and (-1, -5)?

    a. $y = 6x + 11$

    b. $y = 6x$

    c. $y = -6x - 11$

    d. $y = -6x$

    e. $y = 6x - 11$

25. The percentage of smokers above the age of 18 in 2000 was 23.2 percent. The percentage of smokers above the age of 18 in 2015 was 15.1 percent. Find the average rate of change in the percent of smokers above the age of 18 from 2000 to 2015.

    a. -.54 percent

    b. -54 percent

    c. -5.4 percent

    d. -15 percent

    e. -1.5 percent

26. A study of adult drivers finds that it is likely that an adult driver wears his seatbelt. Which of the following could be the probability that an adult driver wears his seat belt?

    a. 0.90

    b. 0.05

    c. 0.25

    d. 0

    e. 1.5

27. In order to estimate deer population in a forest, biologists obtained a sample of deer in that forest and tagged each one of them. The sample had 300 deer in total. They returned a week later and harmlessly captured 400 deer, and 5 were tagged. Use this information to estimate how many total deer were in the forest.

    a. 24,000 deer

    b. 30,000 deer

    c. 40,000 deer

    d. 100,000 deer

    e. 120,000 deer

28. Which of the following is the equation of a vertical line that runs through the point (1, 4)?

    a. $x = 1$

    b. $y = 1$

    c. $x = 4$

    d. $y = 4$

    e. $x = y$

29. What is the missing length x?

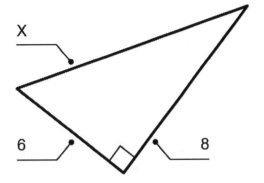

    a. 6
    b. -10
    c. 10
    d. 100
    e. 14

30. What is the correct factorization of the following binomial?
$$2y^3 - 128$$

    a. $2(y + 8)(y - 8)$
    b. $2(y - 4)(y^2 + 4y + 16)$
    c. $2(y + 4)(y - 4)^2$
    d. $2(y - 4)^3$
    e. $2(y - 4)(y^2 + 4y + 16)$

31. What is the simplified form of $(4y^3)^4(3y^7)^2$?
    a. $12y^{26}$
    b. $2304y^{16}$
    c. $12y^{14}$
    d. $2304y^{26}$
    e. $12y^{16}$

32. Use the graph below entitled "Projected Temperatures for Tomorrow's Winter Storm" to answer the question.

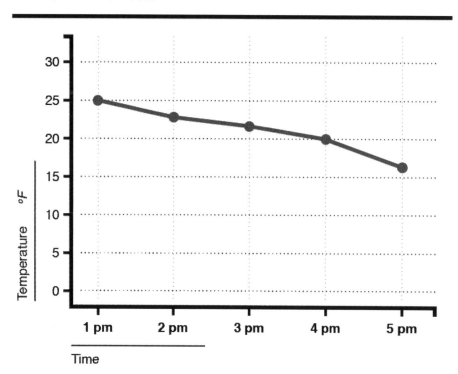

Projected Temperatures for Tomorrow's Winter Storm

What is the expected temperature at 3:00 p.m.?
   a. 25 degrees
   b. 22 degrees
   c. 20 degrees
   d. 16 degrees
   e. 18 degrees

33. The number of members of the House of Representatives varies directly with the total population in a state. If the state of New York has 19,800,000 residents and has 27 total representatives, how many should Ohio have with a population of 11,800,000?
   a. 10
   b. 9
   c. 11
   d. 5
   e. 12

34. Which of these answer choices is a statistical question?
    a. What was your grade on the last test?
    b. What were the grades of the students in your class on the last test?
    c. What kind of car do you drive?
    d. What was Sam's time in the marathon?
    e. What textbooks does Marty use this semester?

Use the following information for questions 35-37.

Eva Jane is practicing for an upcoming 5K run. She has recorded the following times (in minutes):
25, 18, 23, 28, 30, 22.5, 23, 33, 20

35. What is the mean of Eva Jane's time?
    a. 26 minutes
    b. 19 minutes
    c. 24.5 minutes
    d. 23 minutes
    e. 25 minutes

36. What is the mode of Eva Jane's time?
    a. 16 minutes
    b. 20 minutes
    c. 23 minutes
    d. 33 minutes
    e. 25 minutes

37. What is Eva Jane's median score?
    a. 23 minutes
    b. 17 minutes
    c. 28 minutes
    d. 19 minutes
    e. 25 minutes

38. What is the area of the following figure?

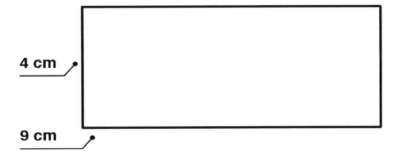

    a. 26 cm
    b. 36 cm
    c. 13 cm^2
    d. 36 cm^2
    e. 65 cm^2

39. What is the volume of the given figure?

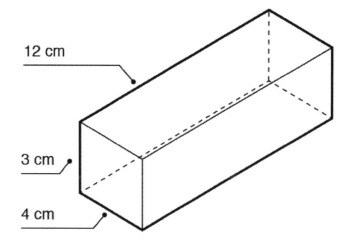

12 cm

3 cm

4 cm

    a. 36 cm²
    b. 144 cm³
    c. 72 cm³
    d. 36 cm³
    e. 144 cm²

40. What type of units are used to describe surface area?
    a. Square
    b. Cubic
    c. Single
    d. Quartic
    e. Volumetric

41. What is the perimeter of the following figure?

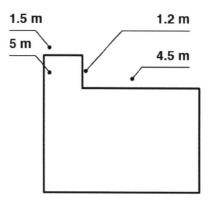

1.5 m   1.2 m

5 m

4.5 m

    a. 13.4 m
    b. 22 m
    c. 12.2 m
    d. 22.5 m
    e. 24.4 m

42. Which equation correctly shows how to find the surface area of a cylinder?

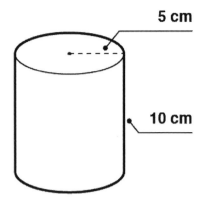

5 cm

10 cm

a. $SA = 2\pi \times 5 \times 10 + 2(\pi 5^2)$
b. $SA = 5 \times 2\pi * 5$
c. $SA = 2\pi 5^2$
d. $SA = 2\pi \times 10 + \pi 5^2$
e. $SA = 2\pi \times 5 \times 10 + \pi 5^2$

43. Which shapes could NOT be used to compose a hexagon?

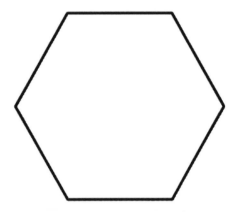

a. Six triangles
b. One rectangle and two triangles
c. Two rectangles
d. Two trapezoids
e. One rectangle and four triangles

Answer Explanations

1. A: First, the distributive property must be used on the left side. This results in $3x + 6 = 14x - 5$. The addition property is then used to add 5 to both sides, and then to subtract $3x$ from both sides, resulting in $11 = 11x$. Finally, the multiplication property is used to divide each side by 11. Therefore, $x = 1$ is the solution.

2. D: First, like terms are collected to obtain $12 - 5x = -5x + 12$. Then, the addition principle is used to move the terms with the variable, so $5x$ is added to both sides and the mathematical statement $12 = 12$ is obtained. This is always true; therefore, all real numbers satisfy the original equation.

3. E: The distributive property is used on both sides to obtain $4x + 20 + 6 = 4x + 6$. Then, like terms are collected on the left, resulting in $4x + 26 = 4x + 6$. Next, the addition principle is used to subtract $4x$ from both sides, and this results in the false statement $26 = 6$. Therefore, there is no solution.

4. D: This is a one-step real-world application problem. The unknown quantity is the number of cases of cola to be purchased. Let x be equal to this amount. Because each case costs $3.50, the total number of cases times $3.50 must equal $40. This translates to the mathematical equation $3.5x = 40$. Divide both sides by 3.5 to obtain $x = 11.4286$, which has been rounded to four decimal places. Because cases are sold whole (the store does not sell portions of cases), and there is not enough money to purchase 12 cases, there is only enough money to purchase 11.

5. A: First, the variables have to be defined. Let x be the first integer; therefore, $x + 1$ is the second integer. This is a two-step problem. The sum of three times the first and two less than the second is translated into the following expression: $3x + (x + 1 - 2)$. This expression is set equal to 411 to obtain $3x + (x + 1 - 2) = 412$. The left-hand side is simplified to obtain $4x - 1 = 411$. The addition and multiplication properties are used to solve for x. First, add 1 to both sides and then divide both sides by 4 to obtain $x = 103$. The next consecutive integer is 104.

6. A: Let x be the unknown, the number of hours Erin can work. We know Katie works $2x$, and the sum of all hours is less than 21. Therefore, $x + 2x < 21$, which simplifies into $3x < 21$. Solving this results in the inequality $x < 7$ after dividing both sides by 3. Therefore, Erin can work less than 7 hours.

7. A: The chart is a bar chart showing how many men and women prefer each genre of movies. The dark gray bars represent the number of women, while the light gray bars represent the number of men. The light gray bars are higher and represent more men than women for the genres of Comedy and Action.

8. E: A line graph represents continuous change over time. The line on the graph is continuous and not broken, as on a scatter plot. Stacked bar graphs are used when comparing multiple variables at one time. They combine some elements of both pie charts and bar graphs, using the organization of bar graphs and the proportionality aspect of pie charts. A bar graph may show change but isn't necessarily continuous over time. A pie graph is better for representing percentages of a whole. Histograms are best used in grouping sets of data in bins to show the frequency of a certain variable.

9. C: The mean for the number of visitors during the first 4 hours is 14. The mean is found by calculating the average for the four hours. Adding up the total number of visitors during those hours gives $12 + 10 + 18 + 16 = 56$. Then, $56 \div 4 = 14$.

10. C: The mode for a set of data is the value that occurs the most. The grade that appears the most is 95. It's the only value that repeats in the set. The mean is around 84.3.

11. B: The relationship between age and time for attention span is a positive correlation because the general trend for the data is up and to the right. As the age increases, so does attention span.

12. A: The area of the shaded region is calculated in a few steps. First, the area of the rectangle is found using the formula $A = length \times width = 6 \times 2 = 12$. Second, the area of the triangle is found using the formula: $A = \frac{1}{2} \times base \times height = \frac{1}{2} \times 3 \times 2 = 3$. The last step is to take the rectangle area and subtract the triangle area. The area of the shaded region is $A = 12 - 3 = 9m^2$.

13. D: The volume for a cylinder is found by using the formula: $V = \pi r^2 h = \pi(2^2) \times 3.5 = 43.96in^3$.

14. C: There are 0.006 kiloliters in 6 liters because 1 liter=0.001kiloliters. The conversion comes from the chart where the prefix kilo- is found three places to the left of the base unit.

15. E: The conversion between feet and centimeters requires a middle term. As there are 2.54 centimeters in 1 inch, the conversion between inches and feet must be found. As there are 12 inches in a foot, the fractions can be set up as follows:

$$3\,feet \times \frac{12\;inches}{1\;foot} \times \frac{2.54\;cm}{1\;inch}$$

The feet and inches cancel out to leave only centimeters for the answer. The numbers are calculated across the top and bottom to yield:

$$\frac{3 \times 12 \times 2.54}{1 \times 1} = 91.44$$

The number and units used together form the answer of 91.44 cm.

16. C: By switching from a radical expression to rational exponents, $\sqrt[4]{x^6} = x^{\frac{6}{4}} = x^{\frac{3}{2}}$. Also, properties of exponents can be used to simplify $\frac{x}{x^3}$ into $x^{1-3} = x^{-2} = \frac{1}{x^2}$. The other terms can be left alone, resulting in an equivalent expression $x^{\frac{3}{2}} - \frac{1}{x^2} + x - 2$.

17. E: Using Descartes' Rule of Signs, count the number of sign changes in coefficients in the polynomial. This results in the number of possible positive zeros. The coefficients are 1, -3, 2, 1, and -3, so the sign changes from 1 to -3, -3 to 2, and 1 to -3, a total of 3 times. Therefore, there are at most 3 positive zeros.

18. A: The corresponding expression written using common denominators of the exponents is $16^{\frac{1}{4}}16^{\frac{2}{4}}$, and then the expression is written as:

$$(16 \times 16^2)^{\frac{1}{4}}$$

This can be written in radical notation as:

$$\sqrt[4]{16^3} = \sqrt[4]{4,096} = \pm 8$$

19. B: The goal is to first isolate the variable. The fractions can easily be cleared by multiplying the entire inequality by 5, resulting in $35 - 4x < 3$. Then, subtract 35 from both sides and divide by -4.

This results in $x > 8$. Notice the inequality symbol has been flipped because both sides were divided by a negative number. The solution set, all real numbers greater than 8, is written in interval notation as $(8, \infty)$. A parenthesis shows that 8 is not included in the solution set.

20. D: Let x be the missing quantity. The problem can be expressed as the following equation:

$$3(5 - x) = x + 5.$$

Distributing the 3 results in:

$$15 - 3x = x + 5$$

Subtract 5 from both sides, add $3x$ to both sides, and then divide both sides by 4. This results in:

$$\frac{10}{4} = \frac{5}{2} = 2.5$$

21. B: The slopes of perpendicular lines are negative reciprocals, meaning their product is equal to -1. The slope of the line given needs to be found. Its equivalent form in slope-intercept form is $y = -\frac{4}{7}x + 23$, so its slope is $-\frac{4}{7}$. The negative reciprocal of this number is $\frac{7}{4}$. The only line in the options given with this same slope is $y = \frac{7}{4}x - 12$.

22. B: The outlier is 35. When a small outlier is removed from a data set, the mean and the median increase. The first step in this process is to identify the outlier, which is the number that lies away from the given set. Once the outlier is identified, the mean and median can be recalculated. The mean will be affected because it averages all of the numbers. The median will be affected because it finds the middle number, which is subject to change because a number is lost. The mode will most likely not change because it is the number that occurs the most, which will not be the outlier if there is only one outlier.

23. A: Division can be used to solve this problem. The division necessary is:

$$\frac{5.972 \times 10^{24}}{7.348 \times 10^{22}}$$

To compute this division, divide the constants first then use algebraic laws of exponents to divide the exponential expression. This results in about 0.8127×10^2, which written in scientific notation is 8.127×10^1.

24. C: First, the slope of the line must be found. This is equal to the change in y over the change in x, given the two points. Therefore, the slope is -6. The slope and one of the points are then plugged into the slope-intercept form of a line: $y - y_1 = m(x - x_1)$. This results in $y - 7 = -6(x + 3)$. The -6 is simplified and the equation is solved for y to obtain $y = -6x - 11$.

25. A: The formula for the rate of change is the same as slope: change in y over change in x. The y-value in this case is percentage of smokers and the x-value is year. The change in percentage of smokers from 2000 to 2015 was 8.1 percent. The change in x was 2000-2015 = -15. Therefore, $\frac{8.1\%}{-15} = -0.54\%$. The percentage of smokers decreased 0.54 percent each year.

26. A: The probability of .9 is closer to 1 than any of the other answers. The closer a probability is to 1, the greater the likelihood that the event will occur. The probability of 0.05 shows that it is very unlikely that an adult driver will wear their seatbelt because it is close to zero. A zero probability means that it will not occur. The probability of 0.25 is closer to zero than to one, so it shows that it is unlikely an adult will wear their seatbelt. Choice *E* is wrong because probability must fall between 0 and 1.

27. A: A proportion should be used to solve this problem. The ratio of tagged to total deer in each instance is set equal, and the unknown quantity is a variable *x*. The proportion is $\frac{300}{x} = \frac{5}{400}$. Cross-multiplying gives $120,000 = 5x$, and dividing through by 5 results in 24,000.

28. A: A vertical line has the same *x* value for any point on the line. Other points on the line would be (1, 3), (1, 5), (1, 9,) etc. Mathematically, this is written as $x = 1$. A vertical line is always of the form $x = a$ for some constant a.

29. C: The Pythagorean Theorem can be used to find the missing length *x* because it is a right triangle. The theorem states that $6^2 + 8^2 = x^2$, which simplifies into $100 = x^2$. Taking the positive square root of both sides results in the missing value $x = 10$.

30. E: First, the common factor 2 can be factored out of both terms, resulting in:

$$2(y^3 - 64)$$

The resulting binomial is a difference of cubes that can be factored using the rule:

$$a^3 - b^3 = (a - b)(a^2 + ab + b^2)$$

a = y and b = 4, therefore, the result is:

$$2(y - 4)(y^2 + 4y + 16)$$

31. D: The exponential rules $(ab)^m = a^m b^m$ and $(a^m)^n = a^{mn}$ can be used to rewrite the expression as $4^4 y^{12} \times 3^2 y^{14}$. The coefficients are multiplied together and the exponential rule $a^m a^n = a^{m+n}$ is then used to obtain the simplified form $2304 y^{26}$.

32. B: Look on the horizontal axis to find 3:00 p.m. Move up from 3:00 p.m. to reach the dot on the graph. Move horizontally to the left to the horizontal axis to between 20 and 25; the best answer choice is 22. The answer of 25 is too high above the projected time on the graph, and the answers of 20 and 16 degrees are too low.

33. B: The number of representatives varies directly with the population, so the equation necessary is $N = k \times P$, where *N* is number of representatives, *k* is the variation constant, and *P* is total population in millions. Plugging in the information for New York allows *k* to be solved for. This process gives $27 = k \times 20$, so $k = 1.35$. Therefore, the formula for number of representatives given total population in millions is $N = 1.35 \times P$. Plugging in $P = 11.6$ for Ohio results in $N = 15.66$, which rounds up to 16 total Representatives.

34. B: This is a statistical question because in order to determine this answer one would need to collect data from each person in the class and it is expected the answers would vary. The other answers do not require data to be collected from multiple sources; therefore, the answers will not vary.

35. E: The mean is found by adding all the times together and dividing by the number of times recorded. $25 + 18 + 23 + 28 + 30 + 22.5 + 23 + 33 + 20 = 222.5$, divided by $9 = 24.7$. Rounding to the nearest minute, the mean is 25 minutes.

36. C: The mode is the time from the data set that occurs most often. The number 23 occurs twice in the data set, while all others occur only once, so the mode is 23.

37. A: To find the median of a data set, you must first list the numbers from smallest to largest, and then find the number in the middle. If there are two numbers in the middle, as in this data set, add the two numbers in the middle together and divide by 2. Putting this list in order from smallest to greatest yields 18, 20, 22.5, 23, 23, 25, 28, 30, and 33, where 23 is the middle number, so 23 minutes is the median.

38. D: The area for a rectangle is found by multiplying the length by the width. The area is also measured in square units, so the correct answer is Choice *D*. The answer of 26 is the perimeter. The answer of 13 is found by adding the two dimensions instead of multiplying.

39. B: The volume of a rectangular prism is found by multiplying the length by the width by the height. This formula yields an answer of 144 cubic units. The answer must be in cubic units because volume involves all three dimensions. Each of the other answers have only two dimensions that are multiplied, and one dimension is forgotten, as in *D*, where 12 and 3 are multiplied, or have incorrect units, as in *E*.

40. A: Surface area is a type of area, which means it is measured in square units. Cubic units are used to describe volume, which has three dimensions multiplied by one another. Quartic units describe measurements multiplied in four dimensions.

41. B: The perimeter is found by adding the length of all the exterior sides. When the given dimensions are added, the perimeter is 22 meters. The equation to find the perimeter can be $P = 5 + 1.5 + 1.2 + 4.5 + 3.8 + 6 = 22$. The last two dimensions can be found by subtracting 1.2 from 5, and adding 1.5 and 4.5, respectively.

42. A: The surface area for a cylinder is the sum of the areas of the two circle bases and the rectangle formed on the side. This is easily seen in the net of a cylinder.

The Net of a Cylinder

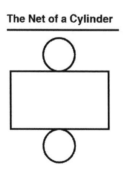

The area of a circle is found by multiplying pi times the radius squared. The rectangle's area is found by multiplying the circumference of the circle by the height.

The equation $SA = 2\pi \times 5 \times 10 + 2(\pi 5^2)$ shows the area of the rectangle as $2\pi \times 5 \times 10$, which yields 314. The area of the bases is found by $\pi 5^2$, which yields 78.5, then multiplied by 2 for the two bases.

43. C: A hexagon can be formed by any combination of the given shapes except for two rectangles. There are no two rectangles that can make up a hexagon.

Reading Comprehension

Summarizing a Complex Text

An important skill is the ability to read a complex text and then reduce its length and complexity by focusing on the key events and details. A **summary** is a shortened version of the original text, written by the reader in their own words. The summary should be shorter than the original text, and it must be thoughtfully formed to include critical points from the original text.

In order to effectively summarize a complex text, it's necessary to understand the original source and identify the major points covered. It may be helpful to outline the original text to get the big picture and avoid getting bogged down in the minor details. For example, a summary wouldn't include a statistic from the original source unless it was the major focus of the text. It's also important for readers to use their own words, yet retain the original meaning of the passage. The key to a good summary is emphasizing the main idea without changing the focus of the original information.

The more complex a text, the more difficult it can be to summarize. Readers must evaluate all points from the original source and then filter out what they feel are the less necessary details. Only the essential ideas should remain. The summary often mirrors the original text's organizational structure. For example, in a problem-solution text structure, the author typically presents readers with a problem and then develops solutions through the course of the text. An effective summary would likely retain this general structure, rephrasing the problem and then reporting the most useful or plausible solutions.

Paraphrasing is somewhat similar to summarizing. It calls for the reader to take a small part of the passage and list or describe its main points. Paraphrasing is more than rewording the original passage, though. Like summary, it should be written in the reader's own words, while still retaining the meaning of the original source. The main difference between summarizing and paraphrasing is that a summary would be appropriate for a much larger text, while paraphrase might focus on just a few lines of text. Effective paraphrasing will indicate an understanding of the original source, yet still help the reader expand on their interpretation. A paraphrase should neither add new information nor remove essential facts that change the meaning of the source.

Inferring the Logical Conclusion from a Reading Selection

Making an inference from a selection means to make an educated guess from the passage read. Inferences should be conclusions based off of sound evidence and reasoning. When multiple-choice test questions ask about the logical conclusion that can be drawn from reading text, the test-taker must identify which choice will unavoidably lead to that conclusion. In order to eliminate the incorrect choices, the test-taker should come up with a hypothetical situation wherein an answer choice is true, but the conclusion is not true. For example, here is an example with three answer choices:

> Fred purchased the newest PC available on the market. Therefore, he purchased the most expensive PC in the computer store.
>
> What can one assume for this conclusion to follow logically?
>
> a. Fred enjoys purchasing expensive items.
> b. PCs are some of the most expensive personal technology products available.
> c. The newest PC is the most expensive one.

The premise of the text is the first sentence: Fred purchased the newest PC. The conclusion is the second sentence: Fred purchased the most expensive PC. Recent release and price are two different factors; the difference between them is the logical gap. To eliminate the gap, one must equate whatever new information the conclusion introduces with the pertinent information the premise has stated. This example simplifies the process by having only one of each: one must equate product recency with product price. Therefore, a possible bridge to the logical gap could be a sentence stating that the newest PCs always cost the most.

Identifying the Topic, Main Idea, and Supporting Details

The **topic** of a text is the general subject matter. Text topics can usually be expressed in one word, or a few words at most. Additionally, readers should ask themselves what point the author is trying to make. This point is the **main idea** of the text, the one thing the author wants readers to know concerning the topic. Once the author has established the main idea, they will support the main idea by supporting details. **Supporting details** are evidence that support the main idea and include personal testimonies, examples, or statistics.

One analogy for these components and their relationships is that a text is like a well-designed house. The topic is the roof, covering all rooms. The main idea is the frame. The supporting details are the various rooms. To identify the topic of a text, readers can ask themselves what or who the author is writing about in the paragraph. To locate the main idea, readers can ask themselves what one idea the author wants readers to know about the topic. To identify supporting details, readers can put the main idea into question form and ask, "what does the author use to prove or explain their main idea?"

Let's look at an example. An author is writing an essay about the Amazon rainforest and trying to convince the audience that more funding should go into protecting the area from deforestation. The author makes the argument stronger by including evidence of the benefits of the rainforest: it provides habitats to a variety of species, it provides much of the earth's oxygen which in turn cleans the atmosphere, and it is the home to medicinal plants that may be the answer to some of the world's deadliest diseases.

Here is an outline of the essay looking at topic, main idea, and supporting details:

Topic: Amazon rainforest
Main Idea: The Amazon rainforest should receive more funding in order to protect it from deforestation.
Supporting Details:
 1. It provides habitats to a variety of species
 2. It provides much of the earth's oxygen which in turn cleans the atmosphere
 3. It is home to medicinal plants that may be the answer to some of the deadliest diseases.

Notice that the topic of the essay is listed in a few key words: "Amazon rainforest." The main idea tells us what about the topic is important: that the topic should be funded in order to prevent deforestation. Finally, the supporting details are what author relies on to convince the audience to act or to believe in the truth of the main idea.

Recognizing Events in a Sequence

Sequence structure is the order of events in which a story or information is presented to the audience. Sometimes the text will be presented in chronological order, or sometimes it will be presented by displaying the most recent information first, then moving backwards in time. The sequence structure depends on the author, the context, and the audience. The structure of a text also depends on the genre in which the text is written. Is it literary fiction? Is it a magazine article? Is it instructions for how to complete a certain task? Different genres will have different purposes for switching up the sequence of their writing.

Narrative Structure

The structure presented in literary fiction is also known as **narrative structure**. Narrative structure is the foundation on which the text moves. The basic ways for moving the text along are in the plot and the setting. The plot is the sequence of events in the narrative that move the text forward through cause and effect. The setting of a story is the place or time period in which the story takes place. Narrative structure has two main categories: linear and nonlinear.

Linear Narrative

Linear narrative is a narrative told in chronological order. Traditional linear narratives will follow the plot diagram below depicting the narrative arc. The narrative arc consists of the exposition, conflict, rising action, climax, falling action, and resolution.

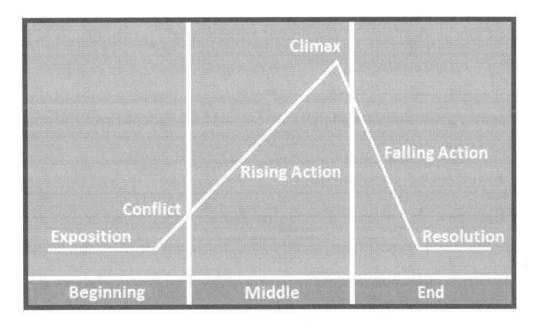

- Exposition: The exposition is in the beginning of a narrative and introduces the characters, setting, and background information of the story. The importance of the exposition lies in its framing of the upcoming narrative. Exposition literally means "a showing forth" in Latin.

- Conflict: The conflict, in a traditional narrative, is presented toward the beginning of the story after the audience becomes familiar with the characters and setting. The conflict is a single instance between characters, nature, or the self, in which the central character is forced to make a decision or move forward with some kind of action. The conflict presents something for the main character, or protagonist, to overcome.

- Rising Action: The rising action is the part of the story that leads into the climax. The rising action will feature the development of characters and plot, and creates the tension and suspense that eventually lead to the climax.

- Climax: The climax is the part of the story where the tension produced in the rising action comes to a culmination. The climax is the peak of the story. In a traditional structure, everything before the climax builds up to it, and everything after the climax falls from it. It is the height of the narrative, and is usually either the most exciting part of the story or is marked by some turning point in the character's journey.

- Falling Action: The falling action happens as a result of the climax. Characters continue to develop, although there is a wrapping up of loose ends here. The falling action leads to the resolution.

- Resolution: The resolution is where the story comes to an end and usually leaves the reader with the satisfaction of knowing what happened within the story and why. However, stories do not always end in this fashion. Sometimes readers can be confused or frustrated at the end from lack of information or the absence of a happy ending.

Nonlinear Narrative

A nonlinear narrative deviates from the traditional narrative in that it does not always follow the traditional plot structure of the narrative arc. Nonlinear narratives may include structures that are disjointed, circular, or disruptive, in the sense that they do not follow chronological order, but rather a nontraditional order of structure. **In medias res** is an example of a structure that predates the linear narrative. *In medias res* is Latin for "in the middle of things," which is how many ancient texts, especially epic poems, began their story, such as Homer's *Iliad*. Instead of having a clear exposition with a full development of characters, they would begin right in the middle of the action.

Modernist texts in the late nineteenth and early twentieth century are known for their experimentation with disjointed narratives, moving away from traditional linear narrative. Disjointed narratives are depicted in novels like *Catch 22*, where the author, Joseph Heller, structures the narrative based on free association of ideas rather than chronology. Another nonlinear narrative can be seen in the novel *Wuthering Heights*, written by Emily Bronte, which disrupts the chronological order by being told retrospectively after the first chapter. There seem to be two narratives in *Wuthering Heights* working at the same time: a present narrative as well as a past narrative. Authors employ disrupting narratives for various reasons; some use it for the purpose of creating situational irony for the readers, while some use it to create a certain effect in the reader, such as excitement, or even a feeling of discomfort or fear.

Sequence Structure in Technical Documents

The purpose of technical documents, such as instructions manuals, cookbooks, or "user-friendly" documents, is to provide information to users as clearly and efficiently as possible. In order to do this, the sequence structure in technical documents that should be used is one that is as straightforward as possible. This usually involves some kind of chronological order or a direct sequence of events. For example, someone who is reading an instruction manual on how to set up their Smart TV wants directions in a clear, simple, straightforward manner that does not leave the reader to guess at the proper sequence or lead to confusion.

Sequence Structure in Informational Texts

The structure in informational texts depends again on the genre. For example, a newspaper article may start by stating an exciting event that happened, and then move on to talk about that event in chronological order, known as **sequence** or **order structure.** Many informational texts also use **cause and effect structure**, which describes an event and then identifies reasons for why that event occurred. Some essays may write about their subjects by way of **comparison and contrast**, which is a structure that compares two things or contrasts them to highlight their differences. Other documents, such as proposals, will have a **problem to solution structure**, where the document highlights some kind of problem and then offers a solution toward the end. Finally, some informational texts are written with lush details and description in order to captivate the audience, allowing them to visualize the information presented to them. This type of structure is known as **descriptive**.

Distinguishing Between Fact and Opinion, Biases, and Stereotypes

Facts and Opinions
A **fact** is a statement that is true empirically or an event that has actually occurred in reality, and can be proven or supported by evidence; it is generally objective. In contrast, an **opinion** is subjective, representing something that someone believes rather than something that exists in the absolute. People's individual understandings, feelings, and perspectives contribute to variations in opinion. Though facts are typically objective in nature, in some instances, a statement of fact may be both factual and yet also subjective. For example, emotions are individual subjective experiences. If an individual says that they feel happy or sad, the feeling is subjective, but the statement is factual; hence, it is a subjective fact. In contrast, if one person tells another that the other is feeling happy or sad—whether this is true or not—that is an assumption or an opinion.

Biases
Biases usually occur when someone allows their personal preferences or ideologies to interfere with what should be an objective decision. In personal situations, someone is biased towards someone if they favor them in an unfair way. In academic writing, being biased in your sources means leaving out objective information that would turn the argument one way or the other. The evidence of bias in academic writing makes the text less credible, so be sure to present all viewpoints when writing, not just your own, so to avoid coming off as biased. Being objective when presenting information or dealing with people usually allows the person to gain more credibility.

Stereotypes
Stereotypes are preconceived notions that place a particular rule or characteristics on an entire group of people. Stereotypes are usually offensive to the group they refer to or allies of that group, and often have negative connotations. The reinforcement of stereotypes isn't always obvious. Sometimes stereotypes can be very subtle and are still widely used in order for people to understand categories within the world. For example, saying that women are more emotional and intuitive than men is a stereotype, although this is still an assumption used by many in order to understand the differences between one another.

Recognizing the Structure of Texts in Various Formats

Text structure is the way in which the author organizes and presents textual information so readers can follow and comprehend it. One kind of text structure is sequence. This means the author arranges the text in a logical order from beginning to middle to end. There are three types of sequences:

- Chronological: ordering events in time from earliest to latest

- Spatial: describing objects, people, or spaces according to their relationships to one another in space

- Order of Importance: addressing topics, characters, or ideas according to how important they are, from either least important to most important

Chronological sequence is the most common sequential text structure. Readers can identify sequential structure by looking for words that signal it, like *first, earlier, meanwhile, next, then, later, finally;* and specific times and dates the author includes as chronological references.

Problem-Solution Text Structure

The problem-solution text structure organizes textual information by presenting readers with a problem and then developing its solution throughout the course of the text. The author may present a variety of alternatives as possible solutions, eliminating each as they are found unsuccessful, or gradually leading up to the ultimate solution. For example, in fiction, an author might write a murder mystery novel and have the character(s) solve it through investigating various clues or character alibis until the killer is identified. In nonfiction, an author writing an essay or book on a real-world problem might discuss various alternatives and explain their disadvantages or why they would not work before identifying the best solution. For scientific research, an author reporting and discussing scientific experiment results would explain why various alternatives failed or succeeded.

Comparison-Contrast Text Structure

Comparison identifies similarities between two or more things. **Contrast** identifies differences between two or more things. Authors typically employ both to illustrate relationships between things by highlighting their commonalities and deviations. For example, a writer might compare Windows and Linux as operating systems, and contrast Linux as free and open-source vs. Windows as proprietary. When writing an essay, sometimes it is useful to create an image of the two objects or events you are comparing or contrasting. Venn diagrams are useful because they show the differences as well as the similarities between two things. Once you've seen the similarities and differences on paper, it might be helpful to create an outline of the essay with both comparison and contrast. Every outline will look different, because every two or more things will have a different number of comparisons and contrasts. Say you are trying to compare and contrast carrots with sweet potatoes.

Here is an example of a compare/contrast outline using those topics:

- Introduction: Share why you are comparing and contrasting the foods. Give the thesis statement.

- Body paragraph 1: Sweet potatoes and carrots are both root vegetables (similarity)

- Body paragraph 2: Sweet potatoes and carrots are both orange (similarity)

- Body paragraph 3: Sweet potatoes and carrots have different nutritional components (difference)

- Conclusion: Restate the purpose of your comparison/contrast essay.

Of course, if there is only one similarity between your topics and two differences, you will want to rearrange your outline. Always tailor your essay to what works best with your topic.

Descriptive Text Structure

Description can be both a type of text structure and a type of text. Some texts are descriptive throughout entire books. For example, a book may describe the geography of a certain country, state, or region, or tell readers all about dolphins by describing many of their characteristics. Many other texts are not descriptive throughout, but use descriptive passages within the overall text. The following are a few examples of descriptive text:

- When the author describes a character in a novel

- When the author sets the scene for an event by describing the setting

- When a biographer describes the personality and behaviors of a real-life individual

- When a historian describes the details of a particular battle within a book about a specific war

- When a travel writer describes the climate, people, foods, and/or customs of a certain place

A hallmark of description is using sensory details, painting a vivid picture so readers can imagine it almost as if they were experiencing it personally.

Cause and Effect Text Structure

When using cause and effect to extrapolate meaning from text, readers must determine the cause when the author only communicates effects. For example, if a description of a child eating an ice cream cone includes details like beads of sweat forming on the child's face and the ice cream dripping down her hand faster than she can lick it off, the reader can infer or conclude it must be hot outside. A useful technique for making such decisions is wording them in "If...then" form, e.g. "If the child is perspiring and the ice cream melting, it may be a hot day." Cause and effect text structures explain why certain events or actions resulted in particular outcomes. For example, an author might describe America's historical large flocks of dodo birds, the fact that gunshots did not startle/frighten dodos, and that because dodos did not flee, settlers killed whole flocks in one hunting session, explaining how the dodo was hunted into extinction.

Interpreting the Meaning of Words and Phrases Using Context

When readers encounter an unfamiliar word in text, they can use the surrounding **context**—the overall subject matter, specific chapter/section topic, and especially the immediate sentence context. Among others, one category of context clues is grammar. For example, the position of a word in a sentence and its relationship to the other words can help the reader establish whether the unfamiliar word is a verb, a noun, an adjective, an adverb, etc. This narrows down the possible meanings of the word to one part of speech. However, this may be insufficient. In a sentence that many birds *migrate* twice yearly, the reader can determine the word is a verb, and probably does not mean eat or drink; but it could mean travel, mate, lay eggs, hatch, molt, etc.

Some words can have a number of different meanings depending on how they are used. For example, the word *fly* has a different meaning in each of the following sentences:

d. "His trousers have a fly on them."
e. "He swatted the fly on his trousers."
f. "Those are some fly trousers."
g. "They went fly fishing."
h. "She hates to fly."
i. "If humans were meant to fly, they would have wings."

As strategies, readers can try substituting a familiar word for an unfamiliar one and see whether it makes sense in the sentence. They can also identify other words in a sentence, offering clues to an unfamiliar word's meaning.

Evaluating the Author's Purpose in a Given Text

Authors may have many purposes for writing a specific text. Their purposes may be to try and convince readers to agree with their position on a subject, to impart information, or to entertain. Other writers are motivated to write from a desire to express their own feelings. Authors' purposes are their reasons for writing something. A single author may have one overriding purpose for writing or multiple reasons. An author may explicitly state their intention in the text, or the reader may need to infer that intention. Those who read reflectively benefit from identifying the purpose because it enables them to analyze information in the text. By knowing why the author wrote the text, readers can glean ideas for how to approach it. The following is a list of questions readers can ask in order to discern an author's purpose for writing a text:

j. From the title of the text, why do you think the author wrote it?
k. Was the purpose of the text to give information to readers?
l. Did the author want to describe an event, issue, or individual?
m. Was it written to express emotions and thoughts?
n. Did the author want to convince readers to consider a particular issue?
o. Was the author primarily motivated to write the text to entertain?
p. Why do you think the author wrote this text from a certain point of view?
q. What is your response to the text as a reader?
r. Did the author state their purpose for writing it?

Students should read to interpret information rather than simply content themselves with roles as text consumers. Being able to identify an author's purpose efficiently improves reading comprehension, develops critical thinking, and makes students more likely to consider issues in depth before accepting writer viewpoints. Authors of fiction frequently write to entertain readers. Another purpose for writing fiction is making a political statement; for example, Jonathan Swift wrote "A Modest Proposal" (1729) as a political satire. Another purpose for writing fiction as well as nonfiction is to persuade readers to take some action or further a particular cause. Fiction authors and poets both frequently write to evoke certain moods; for example, Edgar Allan Poe wrote novels, short stories, and poems that evoke moods of gloom, guilt, terror, and dread. Another purpose of poets is evoking certain emotions: love is popular, as in Shakespeare's sonnets and numerous others. In "The Waste Land" (1922), T.S. Eliot evokes society's alienation, disaffection, sterility, and fragmentation.

Authors seldom directly state their purposes in texts. Some students may be confronted with nonfiction texts such as biographies, histories, magazine and newspaper articles, and instruction manuals, among others. To identify the purpose in nonfiction texts, students can ask the following questions:

s. Is the author trying to teach something?
t. Is the author trying to persuade the reader?
u. Is the author imparting factual information only?
v. Is this a reliable source?
w. Does the author have some kind of hidden agenda?

To apply author purpose in nonfictional passages, students can also analyze sentence structure, word choice, and transitions to answer the aforementioned questions and to make inferences. For example, authors wanting to convince readers to view a topic negatively often choose words with negative connotations.

Narrative Writing

Narrative writing tells a story. The most prominent examples of narrative writing are fictional novels. Here are some examples:

- Mark Twain's The Adventures of Tom Sawyer and The Adventures of Huckleberry Finn

- Victor Hugo's *Les Misérables*

- Charles Dickens' Great Expectations, David Copperfield, and A Tale of Two Cities

- Jane Austen's Northanger Abbey, Mansfield Park, Pride and Prejudice, and Sense and Sensibility

- Toni Morrison's Beloved, The Bluest Eye, and Song of Solomon

- Gabriel García Márquez's One Hundred Years of Solitude and Love in the Time of Cholera

Some nonfiction works are also written in narrative form. For example, some authors choose a narrative style to convey factual information about a topic, such as a specific animal, country, geographic region, and scientific or natural phenomenon.

Since narrative is the type of writing that tells a story, it must be told by someone, who is the narrator. The narrator may be a fictional character telling the story from their own viewpoint. This narrator uses the first person (*I, me, my, mine* and *we, us, our,* and *ours*). The narrator may simply be the author; for example, when Louisa May Alcott writes "Dear reader" in *Little Women*, she (the author) addresses us as readers. In this case, the novel is typically told in third person, referring to the characters as he, she, they, or them. Another more common technique is the omniscient narrator; i.e. the story is told by an unidentified individual who sees and knows everything about the events and characters—not only their externalized actions, but also their internalized feelings and thoughts. Second person, i.e. writing the story by addressing readers as "you" throughout, is less frequently used.

Expository Writing

Expository writing is also known as informational writing. Its purpose is not to tell a story as in narrative writing, to paint a picture as in descriptive writing, or to persuade readers to agree with something as in argumentative writing. Rather, its point is to communicate information to the reader. As such, the point of view of the author will necessarily be more objective. Whereas other types of writing appeal to the reader's emotions, appeal to the reader's reason by using logic, or use subjective descriptions to sway the reader's opinion or thinking, expository writing seeks to do none of these but simply to provide facts, evidence, observations, and objective descriptions of the subject matter. Some examples of expository writing include research reports, journal articles, articles and books about historical events or periods, academic subject textbooks, news articles and other factual journalistic reports, essays, how-to articles, and user instruction manuals.

Technical Writing

Technical writing is similar to expository writing in that it is factual, objective, and intended to provide information to the reader. Indeed, it may even be considered a subcategory of expository writing. However, technical writing differs from expository writing in that (1) it is specific to a particular field, discipline, or subject; and (2) it uses the specific technical terminology that belongs only to that area. Writing that uses technical terms is intended only for an audience familiar with those terms. A primary example of technical writing today is writing related to computer programming and use.

Persuasive Writing

Persuasive writing is intended to persuade the reader to agree with the author's position. It is also known as argumentative writing. Some writers may be responding to other writers' arguments, in which case they make reference to those authors or text and then disagree with them. However, another common technique is for the author to anticipate opposing viewpoints in general, both from other authors and from the author's own readers. The author brings up these opposing viewpoints, and then refutes them before they can even be raised, strengthening the author's argument. Writers persuade readers by appealing to their reason, which Aristotle called **logos**; appealing to emotion, which Aristotle called **pathos**; or appealing to readers based on the author's character and credibility, which Aristotle called **ethos**.

Evaluating the Author's Point of View in a Given Text

When a writer tells a story using the first person, readers can identify this by the use of first-person pronouns, like *I, me, we, us,* etc. However, first-person narratives can be told by different people or from different points of view. For example, some authors write in the first person to tell the story from the main character's viewpoint, as Charles Dickens did in his novels *David Copperfield* and *Great Expectations.* Some authors write in the first person from the viewpoint of a fictional character in the story, but not necessarily the main character. For example, F. Scott Fitzgerald wrote *The Great Gatsby* as narrated by Nick Carraway, a character in the story, about the main characters, Jay Gatsby and Daisy Buchanan. Other authors write in the first person, but as the omniscient narrator—an often-unnamed person who knows all of the characters' inner thoughts and feelings. Writing in first person as oneself is more common in nonfiction.

Third Person

The third-person narrative is probably the most prevalent voice used in fictional literature. While some authors tell stories from the point of view and in the voice of a fictional character using the first person, it is a more common practice to describe the actions, thoughts, and feelings of fictional characters in the third person using *he, him, she, her, they, them,* etc.

Although plot and character development are both necessary and possible when writing narrative from a first-person point of view, they are also more difficult, particularly for new writers and those who find it unnatural or uncomfortable to write from that perspective. Therefore, writing experts advise beginning writers to start out writing in the third person. A big advantage of third-person narration is that the writer can describe the thoughts, feelings, and motivations of every character in a story, which is not possible for the first-person narrator. Third-person narrative can impart information to readers that the characters do not know. On the other hand, beginning writers often regard using the third-person point of view as more difficult because they must write about the feelings and thoughts of every character, rather than only about those of the protagonist.

Second Person

Narrative written in the second person addresses someone else as "you." In novels and other fictional works, the second person is the narrative voice most seldom used. The primary reason for this is that it often reads in an awkward manner, which prevents readers from being drawn into the fictional world of the novel. The second person is more often used in informational text, especially in how-to manuals, guides, and other instructions.

First Person

First person uses pronouns such as *I, me, we, my, us, and our*. Some writers naturally find it easier to tell stories from their own points of view, so writing in the first person offers advantages for them. The first-person voice is better for interpreting the world from a single viewpoint, and for enabling reader immersion in one protagonist's experiences. However, others find it difficult to use the first-person narrative voice. Its disadvantages can include overlooking the emotions of characters, forgetting to include description, producing stilted writing, using too many sentence structures involving "I did....", and not devoting enough attention to the story's "here-and-now" immediacy.

Identifying Primary Sources in Various Media

A **primary sourc**e is a piece of original work, which can include books, musical compositions, recordings, movies, works of visual art (paintings, drawings, photographs), jewelry, pottery, clothing, furniture, and other artifacts. Within books, primary sources may be of any genre. Whether nonfiction based on actual events or a fictional creation, the primary source relates the author's firsthand view of some specific event, phenomenon, character, place, process, ideas, field of study or discipline, or other subject matter. Whereas primary sources are original treatments of their subjects, secondary sources are a step removed from the original subjects; they analyze and interpret primary sources. These include journal articles, newspaper or magazine articles, works of literary criticism, political commentaries, and academic textbooks.

In the field of history, primary sources frequently include documents that were created around the same time period that they were describing, and most often produced by someone who had direct experience or knowledge of the subject matter. In contrast, secondary sources present the ideas and viewpoints of other authors about the primary sources; in history, for example, these can include books and other written works about the particular historical periods or eras in which the primary sources were produced. Primary sources pertinent in history include diaries, letters, statistics, government information, and original journal articles and books. In literature, a primary source might be a literary novel, a poem or book of poems, or a play. Secondary sources addressing primary sources may be criticism, dissertations, theses, and journal articles. Tertiary sources, typically reference works referring to primary and secondary sources, include encyclopedias, bibliographies, handbooks, abstracts, and periodical indexes.

In scientific fields, when scientists conduct laboratory experiments to answer specific research questions and test hypotheses, lab reports and reports of research results constitute examples of primary sources. When researchers produce statistics to support or refute hypotheses, those statistics are primary sources. When a scientist is studying some subject longitudinally or conducting a case study, they may keep a journal or diary. For example, Charles Darwin kept diaries of extensive notes on his studies during sea voyages on the *Beagle*, visits to the Galápagos Islands, etc.; Jean Piaget kept journals of observational notes for case studies of children's learning behaviors. Many scientists, particularly in past centuries, shared and discussed discoveries, questions, and ideas with colleagues through letters, which also constitute primary sources. When a scientist seeks to replicate another's experiment, the reported results, analysis, and commentary on the original work is a secondary source, as is a student's dissertation if it analyzes or discusses others' work rather than reporting original research or ideas.

Comparing and Contrasting Themes from Print and Other Sources

The **theme** of a piece of text is the central idea the author communicates. Whereas the topic of a passage of text may be concrete in nature, the theme is always conceptual. For example, while the topic of Mark Twain's novel *The Adventures of Huckleberry Finn* might be described as something like the coming-of-age experiences of a poor, illiterate, functionally orphaned boy around and on the Mississippi River in 19th-century Missouri, one theme of the book might be that human beings are corrupted by society.

Another might be that slavery and "civilized" society itself are hypocritical. Whereas the main idea in a text is the most important single point that the author wants to make, the theme is the concept or view around which the author centers the text.

Throughout time, humans have told stories with similar themes. Some themes are universal across time, space, and culture. These include themes of the individual as a hero, conflicts of the individual against nature, the individual against society, change vs. tradition, the circle of life, coming-of-age, and the complexities of love. Themes involving war and peace have featured prominently in diverse works, like Homer's *Iliad*, Tolstoy's *War and Peace* (1869), Stephen Crane's *The Red Badge of Courage* (1895), Hemingway's *A Farewell to Arms* (1929), and Margaret Mitchell's *Gone with the Wind* (1936). Another universal literary theme is that of the quest. These appear in folklore from countries and cultures worldwide, including the Gilgamesh Epic, Arthurian legend's Holy Grail quest, Virgil's *Aeneid*, Homer's *Odyssey*, and the *Argonautica*. Cervantes' *Don Quixote* is a parody of chivalric quests. J.R.R. Tolkien's *The Lord of the Rings* trilogy (1954) also features a quest.

One instance of similar themes across cultures is when those cultures are in countries that are geographically close to each other. For example, a folklore story of a rabbit in the moon using a mortar and pestle is shared among China, Japan, Korea, and Thailand—making medicine in China, making rice cakes in Japan and Korea, and hulling rice in Thailand. Another instance is when cultures are more distant geographically, but their languages are related. For example, East Turkestan's Uighurs and people in Turkey share tales of folk hero Effendi Nasreddin Hodja. Another instance, which may either be called cultural diffusion or simply reflect commonalities in the human imagination, involves shared themes among geographically and linguistically different cultures: both Cameroon's and Greece's folklore tell of centaurs; Cameroon, India, Malaysia, Thailand, and Japan, of mermaids; Brazil, Peru, China, Japan, Malaysia, Indonesia, and Cameroon, of underwater civilizations; and China, Japan, Thailand, Vietnam, Malaysia, Brazil, and Peru, of shape-shifters.

Two prevalent literary themes are love and friendship, which can end happily, sadly, or both. William Shakespeare's *Romeo and Juliet*, Emily Brontë's *Wuthering Heights*, Leo Tolstoy's *Anna Karenina*, and both *Pride and Prejudice* and *Sense and Sensibility* by Jane Austen are famous examples. Another theme recurring in popular literature is of revenge, an old theme in dramatic literature, e.g. Elizabethans Thomas Kyd's *The Spanish Tragedy* and Thomas Middleton's *The Revenger's Tragedy*. Some more well-known instances include Shakespeare's tragedies *Hamlet* and *Macbeth*, Alexandre Dumas' *The Count of Monte Cristo*, John Grisham's *A Time to Kill*, and Stieg Larsson's *The Girl Who Kicked the Hornet's Nest*.

Themes are underlying meanings in literature. For example, if a story's main idea is a character succeeding against all odds, the theme is overcoming obstacles. If a story's main idea is one character wanting what another character has, the theme is jealousy. If a story's main idea is a character doing something they were afraid to do, the theme is courage. Themes differ from topics in that a topic is a subject matter; a theme is the author's opinion about it. For example, a work could have a topic of war and a theme that war is a curse. Authors present themes through characters' feelings, thoughts, experiences, dialogue, plot actions, and events. Themes function as "glue" holding other essential story elements together. They offer readers insights into characters' experiences, the author's philosophy, and how the world works.

Practice Questions

Questions 1-6 are based on the following passage from The Life, Crime, and Capture of John Wilkes Booth *by George Alfred Townsend:*

The box in which the President sat consisted of two boxes turned into one, the middle partition being removed, as on all occasions when a state party visited the theater. The box was on a level with the dress circle; about twelve feet above the stage. There were two entrances—the door nearest to the wall having been closed and locked; the door nearest the balustrades of the dress circle, and at right angles with it, being open and left open, after the visitors had entered. The interior was carpeted, lined with crimson paper, and furnished with a sofa covered with crimson velvet, three arm chairs similarly covered, and six cane-bottomed chairs. Festoons of flags hung before the front of the box against a background of lace.

President Lincoln took one of the arm-chairs and seated himself in the front of the box, in the angle nearest the audience, where, partially screened from observation, he had the best view of what was transpiring on the stage. Mrs. Lincoln sat next to him, and Miss Harris in the opposite angle nearest the stage. Major Rathbone sat just behind Mrs. Lincoln and Miss Harris. These four were the only persons in the box.

The play proceeded, although "Our American Cousin," without Mr. Sothern, has, since that gentleman's departure from this country, been justly esteemed a very dull affair. The audience at Ford's, including Mrs. Lincoln, seemed to enjoy it very much. The worthy wife of the President leaned forward, her hand upon her husband's knee, watching every scene in the drama with amused attention. Even across the President's face at intervals swept a smile, robbing it of its habitual sadness.

About the beginning of the second act, the mare, standing in the stable in the rear of the theater, was disturbed in the midst of her meal by the entrance of the young man who had quitted her in the afternoon. It is presumed that she was saddled and bridled with exquisite care.

Having completed these preparations, Mr. Booth entered the theater by the stage door; summoned one of the scene shifters, Mr. John Spangler, emerged through the same door with that individual, leaving the door open, and left the mare in his hands to be held until he (Booth) should return. Booth who was even more fashionably and richly dressed than usual, walked thence around to the front of the theater, and went in. Ascending to the dress circle, he stood for a little time gazing around upon the audience and occasionally upon the stage in his usual graceful manner. He was subsequently observed by Mr. Ford, the proprietor of the theater, to be slowly elbowing his way through the crowd that packed the rear of the dress circle toward the right side, at the extremity of which was the box where Mr. and Mrs. Lincoln and their companions were seated. Mr. Ford casually noticed this as a slightly extraordinary symptom of interest on the part of an actor so familiar with the routine of the theater and the play.

1. Which of the following best describes the author's attitude toward the events leading up to the assassination of President Lincoln?
    a. Excitement, due to the setting and its people
    b. Sadness, due to the death of a beloved president
    c. Anger, due to the impending violence
    d. Neutrality, due to the style of the report
    e. Apprehension, due to the crowd and their ignorance

2. What does the author mean by the last sentence in the passage?
   a. Mr. Ford was suspicious of Booth and assumed he was making his way to Mr. Lincoln's box.
   b. Mr. Ford assumed Booth's movement throughout the theater was due to being familiar with the theater.
   c. Mr. Ford thought that Booth was making his way to the theater lounge to find his companions.
   d. Mr. Ford thought that Booth was elbowing his way to the dressing room to get ready for the play.
   e. Mr. Ford thought that Booth was coming down with an illness due to the strange symptoms he displayed.

3. Given the author's description of the play "Our American Cousin," which one of the following is most analogous to Mr. Sothern's departure from the theater?
   a. A ballet dancer who leaves the New York City Ballet just before they go on to their final performance.
   b. A basketball player leaves an NBA team and the next year they make it to the championship but lose.
   c. A lead singer leaves their band to begin a solo career, and the band's sales on their next album drop by 50 percent.
   d. A movie actor who dies in the middle of making a movie and the movie is made anyway by actors who resemble the deceased.
   e. A professor who switches to the top-rated university for their department only to find the university they left behind has surpassed his new department's rating.

4. Which of the following texts most closely relates to the organizational structure of the passage?
   a. A chronological account in a fiction novel of a woman and a man meeting for the first time.
   b. A cause-and-effect text ruminating on the causes of global warming.
   c. An autobiography that begins with the subject's death and culminates in his birth.
   d. A text focusing on finding a solution to the problem of the Higgs boson particle.
   e. A text contrasting the realities of life on Mars versus life on Earth.

5. Which of the following words, if substituted for the word *festoons* in the first paragraph, would LEAST change the meaning of the sentence?
   a. Feathers
   b. Armies
   c. Adornments
   d. Buckets
   e. Boats

6. What is the primary purpose of the passage?
   a. To persuade the audience that John Wilkes Booth killed Abraham Lincoln
   b. To inform the audience of the setting wherein Lincoln was shot
   c. To narrate the bravery of Lincoln and his last days as President
   d. To recount in detail the events that led up to Abraham Lincoln's death
   e. To disprove the popular opinion that John Wilkes Booth is the person who killed Abraham Lincoln

Questions 7-13 are based on the following passage from The Story of Germ Life *by Herbert William Conn:*

The first and most universal change effected in milk is its souring. So universal is this phenomenon that it is generally regarded as an inevitable change that cannot be avoided, and, as already pointed out, has in the past been regarded as a normal property of milk. To-day, however, the phenomenon is well understood. It is due to the action of certain of the milk bacteria upon the milk sugar which converts it into lactic acid, and this acid gives the sour taste and curdles the milk. After this acid is produced in small quantity its presence proves deleterious to the growth of the bacteria, and further bacterial growth is checked. After souring, therefore, the milk for some time does not ordinarily undergo any further changes.

Milk souring has been commonly regarded as a single phenomenon, alike in all cases. When it was first studied by bacteriologists it was thought to be due in all cases to a single species of micro-organism which was discovered to be commonly present and named *Bacillus acidi lactici*. This bacterium has certainly the power of souring milk rapidly, and is found to be very common in dairies in Europe. As soon as bacteriologists turned their attention more closely to the subject it was found that the spontaneous souring of milk was not always caused by the same species of bacterium. Instead of finding this *Bacillus acidi lactici* always present, they found that quite a number of different species of bacteria have the power of souring milk, and are found in different specimens of soured milk. The number of species of bacteria that have been found to sour milk has increased until something over a hundred are known to have this power. These different species do not affect the milk in the same way. All produce some acid, but they differ in the kind and the amount of acid, and especially in the other changes which are effected at the same time that the milk is soured, so that the resulting soured milk is quite variable. In spite of this variety, however, the most recent work tends to show that the majority of cases of spontaneous souring of milk are produced by bacteria which, though somewhat variable, probably constitute a single species, and are identical with the *Bacillus acidi lactici*. This species, found common in the dairies of Europe, according to recent investigations occurs in this country as well. We may say, then, that while there are many species of bacteria infesting the dairy which can sour the milk, there is one that is more common and more universally found than others, and this is the ordinary cause of milk souring.

When we study more carefully the effect upon the milk of the different species of bacteria found in the dairy, we find that there is a great variety of changes they produce when they are allowed to grow in milk. The dairyman experiences many troubles with his milk. It sometimes curdles without becoming acid. Sometimes it becomes bitter, or acquires an unpleasant "tainted" taste, or, again, a "soapy" taste. Occasionally, a dairyman finds his milk becoming slimy, instead of souring and curdling in the normal fashion. At such times, after a number of hours, the milk becomes so slimy that it can be drawn into long threads. Such an infection proves very troublesome, for many a time it persists in spite of all attempts made to remedy it. Again, in other cases the milk will turn blue, acquiring about the time it becomes sour a beautiful sky-blue colour. Or it may become red, or occasionally yellow. All of these troubles the dairyman owes to the presence in his milk of unusual species of bacteria which grow there abundantly.

7. The word *deleterious* in the first paragraph can be best interpreted as meaning which one of the following?
    a. Amicable
    b. Smoldering
    c. Luminous
    d. Ruinous
    e. Virtuous

8. Which of the following best explains how the passage is organized?

    a. The author begins by presenting the effects of a phenomenon, then explains the process of this phenomenon, and then ends by giving the history of the study of this phenomenon.

    b. The author begins by explaining a process or phenomenon, then gives the history of the study of this phenomenon, this ends by presenting the effects of this phenomenon.

    c. The author begins by giving the history of the study of a certain phenomenon, then explains the process of this phenomenon, then ends by presenting the effects of this phenomenon.

    d. The author begins by giving a broad definition of a subject, then presents more specific cases of the subject, then ends by contrasting two different viewpoints on the subject.

    e. The author begins by contrasting two different viewpoints, then gives a short explanation of a subject, then ends by summarizing what was previously stated in the passage.

9. What is the primary purpose of the passage?

    a. To inform the reader of the phenomenon, investigation, and consequences of milk souring

    b. To persuade the reader that milk souring is due to *Bacillus acidi lactici*, which is commonly found in the dairies of Europe

    c. To describe the accounts and findings of researchers studying the phenomenon of milk souring

    d. To discount the former researchers' opinions on milk souring and bring light to new investigations

    e. To narrate the story of one researcher who discovered the phenomenon of milk souring and its subsequent effects

10. What does the author say about the ordinary cause of milk souring?

    a. Milk souring is caused mostly by a species of bacteria called *Bacillus acidi lactici*, although former research asserted that it was caused by a variety of bacteria.

    b. The ordinary cause of milk souring is unknown to current researchers, although former researchers thought it was due to a species of bacteria called *Bacillus acidi lactici*.

    c. Milk souring is caused mostly by a species of bacteria identical to that of *Bacillus acidi lactici*, although there are a variety of other bacteria that cause milk souring as well.

    d. The ordinary cause of milk souring will sometimes curdle without becoming acidic, though sometimes it will turn colors other than white, or have strange smells or tastes.

    e. The ordinary cause of milk souring is from bacteria with a strange, "soapy" smell, usually the color of sky blue.

11. The author of the passage would most likely agree most with which of the following?

    a. Milk researchers in the past have been incompetent and have sent us on a wild goose chase when determining what causes milk souring.

    b. Dairymen are considered more expert in the field of milk souring than milk researchers.

    c. The study of milk souring has improved throughout the years, as we now understand more of what causes milk souring and what happens afterward.

    d. Any type of bacteria will turn milk sour, so it's best to keep milk in an airtight container while it is being used.

    e. The effects of milk souring is a natural occurrence of milk, so it should not be dangerous to consume.

12. Given the author's account of the consequences of milk souring, which of the following is most closely analogous to the author's description of what happens after milk becomes slimy?
   a. The chemical change that occurs when a firework explodes.
   b. A rainstorm that overwaters a succulent plant.
   c. Mercury inside of a thermometer that leaks out.
   d. A child who swallows flea medication.
   e. A large block of ice that melts into a liquid.

13. What type of paragraph would most likely come after the third?
   a. A paragraph depicting the general effects of bacteria on milk.
   b. A paragraph explaining a broad history of what researchers have found in regard to milk souring.
   c. A paragraph outlining the properties of milk souring and the way in which it occurs.
   d. A paragraph showing the ways bacteria infiltrate milk and ways to avoid this infiltration.
   e. A paragraph naming all the bacteria in alphabetical order with a brief definition of what each does to milk.

Questions 14-20 are based on the following two passages, labeled "Passage A" and "Passage B":

Passage A

(from "Free Speech in War Time" by James Parker Hall, written in 1921, published in Columbia Law Review, Vol. 21 No. 6)

In approaching this problem of interpretation, we may first put out of consideration certain obvious limitations upon the generality of all guaranties of free speech. An occasional unthinking malcontent may urge that the only meaning not fraught with danger to liberty is the literal one that no utterance may be forbidden, no matter what its intent or result; but in fact, it is nowhere seriously argued by anyone whose opinion is entitled to respect that direct and intentional incitations to crime may not be forbidden by the state. If a state may properly forbid murder or robbery or treason, it may also punish those who induce or counsel the commission of such crimes. Any other view makes a mockery of the state's power to declare and punish offences. And what the state may do to prevent the incitement of serious crimes that are universally condemned, it may also do to prevent the incitement of lesser crimes, or of those in regard to the bad tendency of which public opinion is divided. That is, if the state may punish John for burning straw in an alley, it may also constitutionally punish Frank for inciting John to do it, though Frank did so by speech or writing. And if, in 1857, the United States could punish John for helping a fugitive slave to escape, it could also punish Frank for inducing John to do this, even though a large section of public opinion might applaud John and condemn the Fugitive Slave Law.

Passage B

(from "Freedom of Speech in War Time" by Zechariah Chafee, Jr. written in 1919, published in Harvard Law Review Vol. 32 No. 8)

The true boundary line of the First Amendment can be fixed only when Congress and the courts realize that the principle on which speech is classified as lawful or unlawful involves the balancing against each other of two very important social interests, in public safety and in the search for truth. Every reasonable attempt should be made to maintain both interests unimpaired, and the great interest in free speech should be sacrificed only when the interest in public safety is really imperiled, and not, as most men believe, when it is barely conceivable that it may be slightly affected. In war time, therefore, speech should be unrestricted by the censorship or by punishment, unless it is clearly liable to cause direct and dangerous interference with the conduct of the war.

Thus our problem of locating the boundary line of free speech is solved. It is fixed close to the point where words will give rise to unlawful acts. We cannot define the right of free speech with the precision of the Rule against Perpetuities or the Rule in Shelley's Case, because it involves national policies which are much more flexible than private property, but we can establish a workable principle of classification in this method of balancing and this broad test of certain danger. There is a similar balancing in the determination of what is "due process of law." And we can with certitude declare that the First Amendment forbids the punishment of words merely for their injurious tendencies. The history of the Amendment and the political function of free speech corroborate each other and make this conclusion plain.

14. Which one of the following questions is central to both passages?
    a. Why is freedom of speech something to be protected in the first place?
    b. Do people want absolute liberty or do they only want liberty for a certain purpose?
    c. What is the true definition of freedom of speech in a democracy?
    d. How can we find an appropriate boundary of freedom of speech during wartime?
    e. What is the interpretation of the first amendment and its limitations?

15. The authors of the two passages would be most likely to disagree over which of the following?
    a. A man is thrown in jail due to his provocation of violence in Washington D.C. during a riot.
    b. A man is thrown in jail for stealing bread for his starving family, and the judge has mercy for him and lets him go.
    c. A man is thrown in jail for encouraging a riot against the U.S. government for the wartime tactics although no violence ensues.
    d. A man is thrown in jail because he has been caught as a German spy working within the U.S. army.
    e. A man is thrown in jail because he murdered a German-born citizen whom he thought was working for the Central Powers during World War I.

16. The relationship between Passage A and Passage B is most analogous to the relationship between the documents described in which of the following?
    a. A research report that asserts water pollution in major cities in California has increased by thirty percent in the past five years; an article advocating the cessation of chicken farms in California near rivers to avoid pollution.
    b. An article detailing the effects of radiation in Fukushima; a research report describing the deaths and birth defects as a result of the hazardous waste dumped on the Somali Coast.
    c. An article that suggests that labor laws during times of war should be left up to the states; an article that showcases labor laws during the past that have been altered due to the current crisis of war.
    d. A research report arguing that the leading cause of methane emissions in the world is from agriculture practices; an article citing that the leading cause of methane emissions in the world is from the transportation of coal, oil, and natural gas.
    e. A journal article in the Netherlands about the law of euthanasia that cites evidence to support only the act of passive euthanasia as an appropriate way to die; a journal article in the Netherlands about the law of euthanasia that cites evidence to support voluntary euthanasia in any aspect.

17. The author uses the examples in the last lines of Passage A in order to do what?
    a. To demonstrate different types of crimes for the purpose of comparing them to see by which one the principle of freedom of speech would become objectionable.
    b. To demonstrate that anyone who incites a crime, despite the severity or magnitude of the crime, should be held accountable for that crime in some degree.
    c. To prove that the definition of "freedom of speech" is altered depending on what kind of crime is being committed.
    d. To show that some crimes are in the best interest of a nation and should not be punishable if they are proven to prevent harm to others.
    e. To suggest that the crimes mentioned should be reopened in order to punish those who incited the crimes.

18. Which of the following, if true, would most seriously undermine the claim proposed by the author in Passage A that if the state can punish a crime, then it can punish the incitement of that crime?
    a. The idea that human beings are able and likely to change their mind between the utterance and execution of an event that may harm others.
    b. The idea that human beings will always choose what they think is right based on their cultural upbringing.
    c. The idea that the limitation of free speech by the government during wartime will protect the country from any group that causes a threat to that country's freedom.
    d. The idea that those who support freedom of speech probably have intentions of subverting the government.
    e. The idea that if a man encourages a woman to commit a crime and she succeeds, the man is just as guilty as the woman.

19. What is the primary purpose of the second passage?
    a. To analyze the First Amendment in historical situations in order to make an analogy to the current war at hand in the nation.
    b. To demonstrate that the boundaries set during wartime are different from that when the country is at peace, and that we should change our laws accordingly.
    c. To offer the idea that during wartime, the principle of freedom of speech should be limited to that of even minor utterances in relation to a crime.
    d. To call upon the interpretation of freedom of speech to be already evident in the First Amendment and to offer a clear perimeter of the principle during war time.
    e. To assert that any limitation on freedom of speech is a violation of human rights and that the circumstances of war do not change this violation.

20. Which of the following words, if substituted for the word *malecontent* in Passage A, would LEAST change the meaning of the sentence?
    a. Regimen
    b. Cacophony
    c. Anecdote
    d. Residua
    e. Grievance

Questions 21-27 are based on the following passage from Rhetoric and Poetry in the Renaissance: A Study of Rhetorical Terms in English Renaissance Literary Criticism *by DL Clark:*

To the Greeks and Romans, rhetoric meant the theory of oratory. As a pedagogical mechanism, it endeavored to teach students to persuade an audience. The content of rhetoric included all that the ancients had learned to be of value in persuasive public speech. It taught how to work up a case by drawing valid inferences from sound evidence, how to organize this material in the most persuasive order, and how to compose in clear and harmonious sentences. Thus, to the Greeks and Romans, rhetoric was defined by its function of discovering means to persuasion and was taught in the schools as something that every free-born man could and should learn.

In both these respects the ancients felt that poetics, the theory of poetry, was different from rhetoric. As the critical theorists believed that the poets were inspired, they endeavored less to teach men to be poets than to point out the excellences which the poets had attained. Although these critics generally, with the exceptions of Aristotle and Eratosthenes, believed the greatest value of poetry to be in the teaching of morality, no one of them endeavored to define poetry, as they did rhetoric, by its purpose. To Aristotle, and centuries later to Plutarch, the distinguishing mark of poetry was imitation. Not until the renaissance did critics define poetry as an art of imitation endeavoring to inculcate morality . . .

The same essential difference between classical rhetoric and poetics appears in the content of classical poetics. Whereas classical rhetoric deals with speeches which might be delivered to convict or acquit a defendant in the law court, or to secure a certain action by the deliberative assembly, or to adorn an occasion, classical poetic deals with lyric, epic, and drama. It is a commonplace that classical literary critics paid little attention to the lyric. It is less frequently realized that they devoted almost as little space to discussion of metrics. By far the greater bulk of classical treatises on poetics is devoted to characterization and to the technique of plot construction, involving as it does narrative and dramatic unity and movement as distinct from logical unity and movement.

21. What does the author say about one way in which the purpose of poetry changed for later philosophers?
   a. The author says that at first, poetry was not defined by its purpose but was valued for its ability to be used to teach morality. Later, some philosophers would define poetry by its ability to instill morality. Finally, during the renaissance, poetry was believed to be an imitative art, but was not necessarily believed to instill morality in its readers.
   b. The author says that the classical understanding of poetry dealt with its ability to be used to teach morality. Later, philosophers would define poetry by its ability to imitate life. Finally, during the renaissance, poetry was believed to be an imitative art that instilled morality in its readers.
   c. The author says that at first, poetry was thought to be an imitation of reality, then later, philosophers valued poetry more for its ability to instill morality.
   d. The author says that the classical understanding of poetry was that it dealt with the search for truth through its content; later, the purpose of poetry would be through its entertainment value.
   e. The author says that the initial understanding of the purpose of poetry was its entertainment value. Then, as poetry evolved into a more religious era, the renaissance, it was valued for its ability to instill morality through its teaching.

22. What does the author of the passage say about classical literary critics in relation to poetics?
    a. That rhetoric was valued more than poetry because rhetoric had a definitive purpose to persuade an audience, and poetry's wavering purpose made it harder for critics to teach.
    b. That although most poetry was written as lyric, epic, or drama, the critics were most focused on the techniques of lyric and epic and their performance of musicality and structure.
    c. That although most poetry was written as lyric, epic, or drama, the critics were most focused on the techniques of the epic and drama and their performance of structure and character.
    d. That the study of poetics was more pleasurable than the study of rhetoric due to its ability to assuage its audience, and the critics, therefore, focused on what poets did to create that effect.
    e. That since poetics was made by the elite in Greek and Roman society, literary critics resented poetics for its obsession of material things and its superfluous linguistics.

23. What is the primary purpose of this passage?
    a. To alert the readers to Greek and Roman culture regarding poetic texts and the focus on characterization and plot construction rather than lyric and meter.
    b. To inform the readers of the changes in poetic critical theory throughout the years and to contrast those changes to the solidity of rhetoric.
    c. To educate the audience on rhetoric by explaining the historical implications of using rhetoric in the education system.
    d. To convince the audience that poetics is a subset of rhetoric as viewed by the Greek and Roman culture.
    e. To contemplate the differences between classical rhetoric and poetry and to consider their purposes in a particular culture.

24. The word *inculcate* in the second paragraph can be best interpreted as meaning which one of the following?
    a. Imbibe
    b. Instill
    c. Implode
    d. Inquire
    e. Idolize

25. Which of the following most closely resembles the way in which the passage is structured?
    a. The first paragraph presents an issue. The second paragraph offers a solution to the problem. The third paragraph summarizes the first two paragraphs.
    b. The first paragraph presents definitions and examples of a particular subject. The second paragraph presents a second subject in the same way. The third paragraph offers a contrast of the two subjects.
    c. The first paragraph presents an inquiry. The second paragraph explains the details of that inquiry. The last paragraph offers a solution.
    d. The first paragraph presents two subjects alongside definitions and examples. The second paragraph presents us a comparison of the two subjects. The third paragraph presents a contrast of the two subjects.
    e. The first paragraph offers a solution to a problem. The second paragraph questions the solution. The third paragraph offers a different solution.

Answer Explanations

1. D: Neutrality due to the style of the report. The report is mostly objective; we see very little language that entails any strong emotion whatsoever. The story is told almost as an objective documentation of a sequence of actions—we see the president sitting in his box with his wife, their enjoyment of the show, Booth's walk through the crowd to the box, and Ford's consideration of Booth's movements. There is perhaps a small amount of bias when the author mentions the president's "worthy wife." However, the word choice and style show no signs of excitement, sadness, anger, or apprehension from the author's perspective, so the best answer is Choice *D*.

2. B: Mr. Ford assumed Booth's movement throughout the theater was due to being familiar with the theater. Choice *A* is incorrect; although Booth does eventually make his way to Lincoln's box, Mr. Ford does not make this distinction in this part of the passage. Choice *C* is incorrect; although the passage mentions "companions," it mentions Lincoln's companions rather than Booth's companions. Choice *D* is incorrect; the passage mentions "dress circle," which means the first level of the theater, but this is different from a "dressing room." Finally, Choice *E* is incorrect; the passage mentions a "symptom" but does not signify a symptom from an illness.

3. C: A lead singer leaves their band to begin a solo career, and the band's sales on their next album drop by 50 percent. The original source of the analogy displays someone significant to an event who leaves, and then the event becomes the worst for it. We see Mr. Sothern leaving the theater company, and then the play becoming a "very dull affair." Choice *A* depicts a dancer who backs out of an event before the final performance, so this is incorrect. Choice *B* shows a basketball player leaving an event, and then the team makes it to the championship but then loses. This choice could be a contestant for the right answer; however, we don't know if the team has become the worst for his departure or the better for it. We simply do not have enough information here. Choice *D* is incorrect. The actor departs an event, but there is no assessment of the quality of the movie. It simply states what actors filled in instead. Choice *E* is incorrect because the opposite of the source happens; the professor leaves the entity, and the entity becomes better. Additionally, the betterment of the entity is not due to the individual leaving. Choice *E* is not analogous to the source.

4. A: A chronological account in a fiction novel of a woman and a man meeting for the first time. It's tempting to mark Choice A wrong because the genres are different. Choice *A* is a fiction text, and the original passage is not a fictional account. However, the question stem asks specifically for organizational structure. Choice *A* is a chronological structure just like the passage, so this is the correct answer. The passage does not have a cause and effect, problem/solution, or compare/contrast structure, making Choices *B*, *D*, and *E* incorrect. Choice *C* is tempting because it mentions an autobiography; however, the structure of this text starts at the end and works its way toward the beginning, which is the opposite structure of the original passage.

5. C: The word *adornments* would LEAST change the meaning of the sentence because it's the most closely related word to *festoons*. The other choices don't make sense in the context of the sentence. *Feathers* of flags, *armies* of flags, *buckets* of flags, and *boats* of flags are not as accurate as the word *adornments* of flags. The passage also talks about other décor in the setting, so the word *adornments* fits right in with the context of the paragraph.

6. D: The primary purpose of the passage is to recount in detail the events that led up to Abraham Lincoln's death. Choice *A* is incorrect; the author makes no claims and uses no rhetoric of persuasion towards the audience. Choice *B* is incorrect, though it's a tempting choice; the passage depicts the setting in exorbitant detail, but the setting itself is not the primary purpose of the passage. Choice *C* is incorrect; one could argue this is a narrative, and the passage is about Lincoln's last few hours, but this isn't the *best* choice. The best choice recounts the details that leads up to Lincoln's death. Finally, Choice *E* is incorrect. The author does not try to prove or disprove anything to the audience, and the passage does not even make it to when Lincoln gets shot, so this part of the story is irrelevant.

7. D: The word *deleterious* can be best interpreted as referring to the word *ruinous*. The first paragraph attempts to explain the process of milk souring, so the "acid" would probably prove "ruinous" to the growth of bacteria and cause souring. Choice *A*, *amicable*, means friendly, so this does not make sense in context. Choice *B*, *smoldering*, means to boil or simmer, so this is also incorrect. Choices *C* and *E*, *luminous* and *virtuous*, have positive connotations and don't make sense in the context of the passage. Luminous means shining or brilliant, and virtuous means to be honest or ethical.

8. B: The author begins by explaining a process or phenomenon, then gives the history of the study of this phenomenon, and ends by presenting the effects of this phenomenon. The author explains the process of souring in the first paragraph by informing the reader that "it is due to the action of certain of the milk bacteria upon the milk sugar which converts it into lactic acid, and this acid gives the sour taste and curdles the milk." In the second paragraph, we see how the phenomenon of milk souring was viewed when it was "first studied," and then we proceed to gain insight into "recent investigations" toward the end of the paragraph. Finally, the passage ends by presenting the effects of the phenomenon of milk souring. We see the milk curdling, becoming bitter, tasting soapy, turning blue, or becoming thread-like. All of the other answer choices are incorrect.

9: A: The primary purpose is to inform the reader of the phenomenon, investigation, and consequences of milk souring. Choice *B* is incorrect because the passage states that *Bacillus acidi lactici* is not the only cause of milk souring. Choice *C* is incorrect because, although the author mentions the findings of researchers, the main purpose of the text does not seek to describe their accounts and findings, as we are not even told the names of any of the researchers. Choice *D* is tricky. We do see the author present us with new findings in contrast to the first cases studied by researchers. However, this information is only in the second paragraph, so it is not the primary purpose of the *entire passage*. Finally, Choice *E* is incorrect because the genre of the passage is more informative than narrative, although the author does talk about the phenomenon of milk souring and its subsequent effects.

10. C: Milk souring is caused mostly by a species of bacteria identical to that of *Bacillus acidi lactici* although there are a variety of other bacteria that cause milk souring as well. Choice *A* is incorrect because it contradicts the assertion that the souring is still caused by a variety of bacteria. Choice *B* is incorrect because the ordinary cause of milk souring *is known* to current researchers. Choice *D* is incorrect because this names mostly the effects of milk souring, not the cause. Choice *E* is incorrect because the bacteria itself doesn't have a strange soapy smell or is a different color, but it eventually will cause the milk to produce these effects.

11. C: The study of milk souring has improved throughout the years, as we now understand more of what causes milk souring and what happens afterward. None of the choices here are explicitly stated, so we have to rely on our ability to make inferences. Choice *A* is incorrect because there is no indication from the author that milk researchers in the past have been incompetent—only that recent research has done a better job of studying the phenomenon of milk souring. Choice *B* is incorrect because the author refers to dairymen in relation to the effects of milk souring and their "troubles" surrounding milk souring, and does not compare them to milk researchers. Choice *D* is incorrect because we are told in the second paragraph that only certain types of bacteria are able to sour milk. Choice *E* is incorrect; although we are told that milk souring is a natural occurrence, the author makes no implication that soured milk is safe to consume. Choice *C* is the best answer choice here because although the author does not directly state that the study of milk souring has improved, we can see this might be true due to the comparison of old studies to newer studies, and the fact that the newer studies are being used as a reference in the passage.

12. A: It is most analogous to the chemical change that occurs when a firework explodes. The author tells us that after milk becomes slimy, "it persists in spite of all attempts made to remedy it," which means the milk has gone through a chemical change. It has changed its state from milk to sour milk by changing its odor, color, and material. After a firework explodes, there is nothing one can do to change the substance of a firework back to its original form—the original substance is turned into sound and light. Choice *B* is incorrect because, although the rain overwatered the plant, it's possible that the plant is able to recover from this. Choice *C* is incorrect because although mercury leaking out may be dangerous, the actual substance itself stays the same and does not alter into something else. Choice *D* is incorrect; this situation is not analogous to the alteration of a substance. Choice *E* is also incorrect. Ice melting into a liquid is a physical change, which means it can be undone. Milk turning sour, as the author asserts, cannot be undone.

13. D: It would most likely be a paragraph showing the ways bacteria infiltrate milk and ways to avoid this infiltration. Choices *A, B,* and *C* are incorrect because these are already represented in the third, second, and first paragraphs. Choice *E* is incorrect; this choice isn't impossible. There could be a glossary right after the third paragraph, but this would be an awkward place for a glossary. Choice *D* is the best answer because it follows a sort of problem/solution structure in writing.

14. E: A central question to both passages is: What is the interpretation of the first amendment and its limitations? Choice *A* is incorrect; this is a question for the first passage but it does not apply to the second. Choice *B* is incorrect; a quote mentions this at the end of the first passage, but this question is not found in the second passage. Choice *C* is incorrect, as the passages are not concerned with the definition of freedom of speech, but how to interpret it. Choice *D* is incorrect; this is a question for the second passage, but it is not found in the first passage.

15. C: The authors would most likely disagree over the man thrown in jail for encouraging a riot against the U.S. government for the wartime tactics although no violence ensued. The author of Passage A says that "If a state may properly forbid murder or robbery or treason, it may also punish those who induce or counsel the commission of such crimes." This statement tells us that the author of Passage A would support throwing the man in jail for encouraging a riot, although no violence ensues. The author of Passage B states that "And we can with certitude declare that the First Amendment forbids the punishment of words merely for their injurious tendencies." This is the best answer choice because we are clear on each author's stance in this situation. Choice *A* is tricky; the author of Passage A would definitely agree with this, but it's questionable whether the author of Passage B would also agree. Violence does ensue at the capitol as a result of this man's provocation, and the author of Passage B states "speech should be unrestricted by censorship . . . unless it is clearly liable to cause direct . . . interference with the conduct of war." This answer is close, but it is not the *best* choice. Choice *B* is incorrect because we have no way of knowing what the authors' philosophies are in this situation. Choice *D* is incorrect because, again, we have no way of knowing what the authors would do in this situation, although it's assumed they would probably both agree with this. Choice *E* is something the authors would probably both agree on, because brutal violence ensued, but it has nothing to do with free speech, so we have no way of knowing for sure.

16. E: Choice *E* is the best answer. To figure out the correct answer choice we must find out the relationship between Passage A and Passage B. Between the two passages, we have a general principle (freedom of speech) that is questioned on the basis of interpretation. In Choice *E*, we see that we have a general principle (right to die, or euthanasia) that is questioned on the basis of interpretation as well. Should euthanasia only include passive euthanasia, or euthanasia in any aspect? Choice *A* is a problem/solution relationship; the first option outlines a problem, and the second option delivers a solution, so this choice is incorrect. Choice *B* is incorrect because it does not question the interpretation of a principle, but rather describes the effects of two events that happened in the past involving contamination of radioactive substances. Choice *C* begins with a principle—that of labor laws during wartime—but in the second option, the interpretation isn't questioned. The second option looks at the historical precedent of labor laws in the past during wartime. Choice *D* is incorrect because the two texts disagree over the cause of something rather than the interpretation of it.

17. B: Choice *B* is the best answer choice because the author is trying to demonstrate via the examples that anyone who incites a crime, despite the severity or magnitude of the crime, should be held accountable for that crime in some degree. Choice *A* is incorrect because the crimes mentioned are not being compared to each other, but they are being used to demonstrate a point. Choice *C* is incorrect because the author makes the same point using both of the examples and does not question the definition of freedom of speech but its ability to be limited. Choice *D* is incorrect because this sentiment goes against what the author has been arguing throughout the passage. Choice *E* is incorrect because the author does not suggest that the crimes mentioned be reopened anywhere in the passage.

18. A: The idea that human beings are able and likely to change their mind between the utterance and execution of an event that may harm others most seriously undermines the claim because it brings into question the bad tendency of a crime and points out the difference between utterance and action in moral situations. Choice *B* is incorrect; this idea does not undermine the claim at hand, but introduces an observation irrelevant to the claim. Choices *C, D,* and *E* would most likely strengthen the argument's claim; or, they are at least supported by the author in Passage A.

19. D: The primary purpose is to call upon the interpretation of freedom of speech to be already evident in the First Amendment and to offer a clear perimeter of the principle during war time. Choice *A* is incorrect; the passage calls upon no historical situations as precedent in this passage. Choice *B* is incorrect; we can infer that the author would not agree with this, because the author states that "In war time, therefore, speech should be unrestricted . . . by punishment." Choice *C* is incorrect; this is more consistent with the main idea of the first passage. Choice *E* is incorrect; the passage states a limitation in saying that "speech should be unrestricted . . . unless it is clearly liable to cause direct and dangerous interference with the conduct of war."

20. E: The word that would least change the meaning of the sentence is *grievance*. *Malcontent* is a complaint or grievance, and in this context would be uttered in advocation of absolute freedom of speech. Choice *A*, *regimen*, means a pattern of living, and would not make sense in this context. Choice *B*, *cacophony*, means a harsh noise or mix of discordant noises; someone may express or "urge" a cacophony but it would be an awkward word in this context. Choice *C*, *anecdote*, is a short account of an amusing story. Since the word is a noun, it fits grammatically inside the sentence, but anecdotes are usually thought out, and this word is considered "unthinking." Choice *D*, *residua*, means an outcome, and also does not make sense within this context.

21. B: The author says that the classical understanding of poetry dealt with its ability to be used to teach morality. Later, philosophers would define poetry by its ability to imitate life. Finally, during the renaissance, poetry was believed to be an imitative art that instilled morality in its readers. The rest of the answer choices improperly interpret this explanation in the passage. Poetry was never mentioned for use in entertainment, which makes Choices *D* and *E* incorrect. Choices *A* and *C* are incorrect because they mix up the chronological order.

22. C: The author says that although most poetry was written as lyric, epic, or drama, the critics were most focused on the techniques of the epic and drama and their performance of structure and character. This is the best answer choice as portrayed by paragraph three. Choice *A* is incorrect because nowhere in the passage does it say rhetoric was more valued than poetry, although it did seem to have a more definitive purpose than poetry. Choice *B* is incorrect; this almost mirrors Choice *A*, but the critics were *not* focused on the lyric, as the passage indicates. Choice *D* is incorrect because the passage does not mention that the study of poetics was more pleasurable than the study of rhetoric. Choice *E* is incorrect because again, we do not see anywhere in the passage where poetry was reserved for the most elite in society.

23. E: The purpose is to contemplate the differences between classical rhetoric and poetry and to consider their purposes in a particular culture. Choice *A* is incorrect; this thought is discussed in the third paragraph, but it is not the main idea of the passage. Choice *B* is incorrect; although changes in poetics throughout the years is mentioned, this is not the main idea of the passage. Choice *C* is incorrect; although this is partly true—that rhetoric within the education system is mentioned—the subject of poetics is left out of this answer choice. Choice *D* is incorrect; the passage makes no mention of poetics being a subset of rhetoric.

24. B: The correct answer choice is Choice *B*, *instill*. Choice *A*, *imbibe*, means to drink heavily, so this choice is incorrect. Choice *C*, *implode*, means to collapse inward, which does not make sense in this context. Choice *D*, *inquire*, means to investigate. This option is better than the other options, but it is not as accurate as *instill*. Choice *E*, *idolize*, means to admire, which does not make sense in this context.

25. B: The first paragraph presents definitions and examples of a particular subject. The second paragraph presents a second subject in the same way. The third paragraph offers a contrast of the two subjects. In the passage, we see the first paragraph defining rhetoric and offering examples of how the Greeks and Romans taught this subject. In the second paragraph, poetics is defined and examples of its dynamic definition are provided. In the third paragraph, the contrast between rhetoric and poetry is characterized through how each of these were studied in a classical context.

Situational Judgment

The Situational Judgment section of the AFOQT measures the judgment you'd exercise when dealing with subordinates and supervisors, especially when there's an interpersonal conflict or ethical dilemma. This section is composed of fifty questions, and each question presents a different hypothetical situation. After the situation, there will be five possible responses and two questions asking which of the following responses is "most effective" and which of the following responses is "least effective."

Correct answers are determined by a consensus of experienced U.S. Air Force Officers. The best answers adhere to an honest and ethical standard without creating additional problems. Use your best analytical reasoning here. One tip would be to imagine how each response would play out in real life. What would the consequences of this response be? Would it help or hurt other members of the team? Is it honest and ethical? Is it necessary? Does it create additional problems? Although some questions may seem like they have two appropriate responses, only one can be selected, so it's important to consider which answer is best, not just appropriate.

Practice Questions

Situation 1

A new shipment of twelve boxes has been delivered into your unit, and they're stacked up in the supply room. You look at the incoming shipment spreadsheet, and you notice that six of the twelve boxes haven't been scanned into the system. One of your subordinates is in charge of incoming shipments, while you're the leader of the entire unit. It's a priority that the entire shipment is scanned in as soon as it arrives at the warehouse.

Possible actions:
  a. Scan the shipment into the system yourself, and tell a senior officer that you're having problems with one of your subordinates.
  b. Scan the shipment in yourself, and say nothing to the subordinate.
  c. Remind your subordinate of their responsibilities with incoming shipments, and ask them to scan the rest of the shipment into the system.
  d. Write a note to your subordinate asking them to scan the boxes into the system when they have the time.
  e. Criticize your subordinate for not doing their job properly, and tell them they'll be moved to a different unit if they forget again.

1. Select the MOST EFFECTIVE action (a – e) in response to the situation.

2. Select the LEAST EFFECTIVE action (a – e) in response to the situation.

Situation 2

You and a coworker each lead separate teams that are now working on the same assignment. The success of the assignment depends upon teamwork since both teams need to simultaneously produce quality work for their assigned tasks. Your coworker has just come to you and explained that two of their team members are out of the assignment for various reasons. The coworker tells you that they need some extra help from your team to accomplish their team's tasks.

Possible actions:

  a. Refuse to help the coworker since it's their fault they put a bad team together.
  b. Take on whatever work the coworker needs done, and spread it out amongst your own team.
  c. Request a meeting with the senior officer to determine the appropriate action to take.
  d. Offer to help the coworker by splitting the extra work between both teams in an appropriate manner.
  e. Tell the coworker you're sorry, but you don't think it's right that your team should suffer for someone else's mistake.

3. Select the MOST EFFECTIVE action (a – e) in response to the situation.

4. Select the LEAST EFFECTIVE action (a – e) in response to the situation.

174

Situation 3

You notice that you're feeling fatigued by your workload, and your quality of work is declining. You're frustrated with your unit and what's constantly being asked of you. You notice that your negative attitude is starting to affect the team, and they walk on eggshells around you most of the time. You decide that you want to transfer into a different unit to see if that'll help with the burnout.

Possible actions:

a. Meet with a senior officer to let them know about your recent frustrations and your desire to change units.
b. Talk to your coworkers about your behavior, and ask for their advice on what you should do.
c. Don't do anything; you should deal with the frustration on your own.
d. Pretend nothing is wrong, act more courteous to your team, and do everything to make them believe that you're satisfied with your work.
e. Start communicating with another unit to express your desire to join them, and begin making preparations to leave.

5. Select the MOST EFFECTIVE action (a – e) in response to the situation.

6. Select the LEAST EFFECTIVE action (a – e) in response to the situation.

Situation 4

Over the last few months you've noticed that your unit's organizational structure is ineffective in key areas. You come up with a new organizational structure that directly addresses those problems, and you believe that it's much more efficient than the one already in place. You mention this new organizational structure to your coworkers, and they agree with your proposal. However, the senior officer is vehemently opposed to the idea, as they believe it will take months to implement.

Possible actions:

a. Since the senior officer already gave their opinion on the matter, you should drop the matter and accept the status quo.
b. Create a presentation of your proposed organizational structure to show how it will benefit the unit in the long run, ask permission to present it to the senior officer, and accept their final decision of yes or no.
c. Go ahead and begin to implement your new organizational structure regardless of the senior officer's opinion, as you know this will benefit the unit as a whole.
d. Continue working within the current organizational structure, but keep a list of all the ways your structure would've fixed problems your unit experiences.
e. Confront the senior officer, and tell him that everyone supports your decision to move forward with the new organizational structure since it's the best plan.

7. Select the MOST EFFECTIVE action (a – e) in response to the situation.

8. Select the LEAST EFFECTIVE action (a – e) in response to the situation.

Situation 5

You and a coworker notice that a file marked "confidential" has been taken out of the main office. You realize that you were supposed to file this folder the day before, but you were distracted by an emergency and forgot about it. The file contains information that is above the security clearance of your subordinates who also work in the same office.

Possible actions:

a. Lock down the building, and tell the senior officer that one of your subordinates stole the file.
b. Inform your coworker that you have no idea what happened, and ask them to report the missing file to the senior officer.
c. Gather your subordinates together to help you find the missing file.
d. Search the building immediately for the missing file by yourself.
e. Notify the senior officer of your mistake right away, and ask what the best course of action to take is.

9. Select the MOST EFFECTIVE action (a – e) in response to the situation.

10. Select the LEAST EFFECTIVE action (a – e) in response to the situation.

Situation 6

While in a meeting, your superior officer and their general officer get into an argument about whether or not to implement a new program in the unit. You work with your superior officer every day and rarely see their general officer, but the general officer outranks both of you. You know that these two have a history of personal conflict. After the argument ends, they ask for your opinion on what should be done about the program.

Possible actions:

a. Take the side of the general officer since the general officer has more power than your superior officer, and you'd have a better chance for a promotion if you took their side.
b. Take the side of your superior officer since you have to see them every day, and your job would be miserable if you made them angry.
c. Weigh the advantages and disadvantages of each side, and give your honest opinion on the subject without becoming involved in their personal conflict.
d. Refuse to answer, because you don't want to become involved in their personal conflict.
e. Tell them your opinion, and try to solve the personal conflict between them.

11. Select the MOST EFFECTIVE action (a – e) in response to the situation.

12. Select the LEAST EFFECTIVE action (a – e) in response to the situation.

Situation 7

As the leader of your unit, you've always had a hard-working team and have achieved excellent results in your work. In the past few months, you've let personal issues affect your work, which has negatively impacted your team as a whole. In addition, budget cuts have left your team without the necessary resources to complete your projects on time. A senior officer confronts you about your team's poor performance and doesn't take the budget cuts into consideration, blaming only you for the poor performance.

Possible actions:

a. Put aside your personal issue while at work, and contact other officers to see if they have experience to offer when dealing with budget cuts.
b. Explain to the senior officer that the budget cuts are the real reason why your team is performing poorly.
c. Let the senior officer know that you need some time to deal with your personal issue and apologize for your poor performance.
d. Go to the senior officer's superior, tell them that you're being treated unfairly, and explain the budget cuts to them.
e. Tell the senior officer that your team is not pulling their weight, and criticize them for not working hard enough to overcome the budget cuts.

13. Select the MOST EFFECTIVE action (a – e) in response to the situation.

14. Select the LEAST EFFECTIVE action (a – e) in response to the situation.

Situation 8

Recently, a new officer has moved into your unit as one of your subordinates. You notice that they're very popular with the other officers and also with your superior officer. You notice that they start undermining your decisions, and some of your fellow officers have mentioned that the new subordinate has expressed a desire to take over some of your job responsibilities.

Possible actions:

a. Let the rest of the unit know that this subordinate is being uncooperative to shame them into changing their behavior.
b. Approach your subordinate, let them know how important it is for the unit to follow the chain-of-command, and explain why cooperation is essential if you're going to work together.
c. Go to your superior officer, and ask them to tell your subordinate that their behavior is unacceptable.
d. Do nothing; these situations tend to take care of themselves.
e. Write a letter to your subordinate detailing how his behavior is negatively impacting the unit.

15. Select the MOST EFFECTIVE action (a – e) in response to the situation.

16. Select the LEAST EFFECTIVE action (a – e) in response to the situation.

Situation 9

You notice over the past month that one of your subordinates has developed a negative attitude towards their work assignments and frequently misses important deadlines. You've never had this problem with them before. In fact, they're usually one of your best people, and the quality of their work is typically excellent.

Possible actions:

a. Write a letter to your superior officer, and ask for their advice on how to handle the situation.
b. Let your subordinate know that they'll be terminated if their behavior doesn't change, and implement a strict action plan with clearly defined rules to follow.
c. Meet with your subordinate to tell them their behavior is negatively affecting the unit, and ask if there's anything you can do to help them better perform their role.
d. Ask the unit as a whole to confront your subordinate so that they realize the consequences of their negative behavior.
e. Don't do anything. Your subordinate's attitude will probably change eventually, and you don't want to lose a valuable team member because of a single rough patch.

17. Select the MOST EFFECTIVE action (a – e) in response to the situation.

18. Select the LEAST EFFECTIVE action (a – e) in response to the situation.

Situation 10

You've just joined a new unit, and the commanding officer has been an effective mentor. You've learned many new things, and you feel that you can work independently without any additional assistance. However, you feel that your superior officer is not giving you the space you need.

Possible actions:

a. Start making a list of all the tasks you can effectively do by yourself, and show it to your superior officer as proof that you're ready for more independence.
b. Avoid your superior officer in the hopes that they'll back off after realizing that you're doing the job perfectly fine without their help.
c. Politely tell your superior officer that their decision to hover over you is preventing you from doing your best work.
d. Tell your coworkers about your desire for more independence, and hope that one of them relays the message to your superior officer.
e. Meet with your superior officer, thank them for their guidance, and then open a dialogue about having a more independent role.

19. Select the MOST EFFECTIVE action (a – e) in response to the situation.

20. Select the LEAST EFFECTIVE action (a – e) in response to the situation.

Situation 11

You and a fellow officer have worked on the same assignment for the past few months. Though you've split the work evenly, you find yourself doing more than your fair share. The other officer has been out sick for about half the time you've been working together, and when they're in the office, they're constantly getting frustrated by being so behind on the project.

Possible actions:

a. Take on the additional workload without complaining.
b. Go straight to your superior officer, and tell them that your partner is not pulling their weight.
c. Write an email to your partner when they're out sick to explain your frustration with their absence.
d. Let your partner know the current situation is not working out, and request guidance from your superior officer on what to do.
e. Do nothing; the project might turn out fine in the end.

21. Select the MOST EFFECTIVE action (a – e) in response to the situation.

22. Select the LEAST EFFECTIVE action (a – e) in response to the situation.

Situation 12

You're interviewing two candidates for a position in your unit. The first candidate is a perfect fit for the position, and you can tell you'll get along great with them. The second candidate is good friends with your superior officer, and they're also a good fit for the position, though probably not as qualified as the first candidate. Your superior officer asks for your opinion about which candidate they should hire.

Possible actions:

a. Tell your superior officer that you don't have an opinion to avoid the drama of favoritism.
b. Refuse to answer the question, because you think it's unfair that you have to give your opinion on the matter after your superior officer already made their decision.
c. Tell your superior officer that candidate two is more qualified since you know this is the candidate your superior officer wants to hire.
d. Tell your superior officer that you'd like new candidates to choose from since deciding between these two candidates creates a conflict of interest.
e. Tell your superior officer that the first candidate is more qualified for this position.

23. Select the MOST EFFECTIVE action (a – e) in response to the situation.

24. Select the LEAST EFFECTIVE action (a – e) in response to the situation.

Situation 13

New equipment has just arrived for your team. This equipment is desperately needed, as what you've been using lately is falling apart. Your superior officer reads off the names on the checklist, and everyone receives their new equipment except for you.

Possible actions:

a. Go to a subordinate, and demand that they give you their new equipment since you need it more than they do.
b. Complain to the department that sent over the new equipment, and ask them to send you the equipment that you need to do your job.
c. Let your superior officer know right away that there's been a mistake, and you didn't receive the updated equipment that you need to do your job.
d. Confront your superior officer, and demand to receive new equipment like everyone else.
e. Tell your superior officer that you're going to transfer to a new unit that provides you with the equipment that's necessary to do your job.

25. Select the MOST EFFECTIVE action (a – e) in response to the situation.

26. Select the LEAST EFFECTIVE action (a – e) in response to the situation.

Situation 14

You've been keeping inventory in the supply room for the past couple of weeks. You keep track of everyone coming in and out, and you also keep records of the most important inventory still remaining after every visit. You notice that when one particular fellow officer visits there's always something valuable missing. This past week, after the officer left, a small gold statue honoring a previous Air Force Officer was gone.

Possible actions:

a. Wait until you can catch the officer in the act, and then accuse them of stealing.
b. Do nothing. You don't want to accuse anyone, and eventually, it will stop.
c. Go to your superior officer, and tell them a fellow officer has been stealing from the supply room.
d. Approach the officer, let them know what you've discovered, and ask if they can offer an explanation.
e. Tell your coworkers about the situation, and ask them to watch the fellow officer for any suspicious behavior.

27. Select the MOST EFFECTIVE action (a – e) in response to the situation.

28. Select the LEAST EFFECTIVE action (a – e) in response to the situation.

Situation 15

As you were sending an email around noon, you realized you accidentally sent the message to someone who doesn't have the appropriate clearance to see the confidential information attached to the email.

Possible actions:

a. Tell your superior officer about it, and ask that they deal with the mistake.
b. Don't do anything, and hope that nobody notices what happened.
c. Contact the person you wrongly emailed, ask them to delete it, and then let your superior officer know what happened.
d. Claim you delegated the task to your subordinate, and accuse them of sending the email.
e. Ask your coworkers for advice, and do whatever the majority thinks.

29. Select the MOST EFFECTIVE action (a – e) in response to the situation.

30. Select the LEAST EFFECTIVE action (a – e) in response to the situation.

Situation 16

You've recently started working at a new base, and you're at a meeting concerning the training of new recruits for the upcoming year. The meeting is with senior officers, and you're their subordinate. You were invited to this meeting as a courtesy, so that you can get to know the base and its hierarchy. The senior officers are having an argument about whether or not they should change some of the routines created for new recruit training. One of the senior officers asks you to give your opinion on the subject.

Possible actions:

a. Explain your rank to the senior officer, and admit that you're not experienced with the subject before giving your opinion.
b. Tell the senior officer that you believe his decision on the matter seems respectable, and he should have the final say, despite whether or not you agree with him.
c. Ask the senior officer if you could give your opinion to him after the meeting in a more private setting.
d. Ask the person that you came with to answer in your place.
e. Don't respond to the senior officer since you don't have the experience to answer the question.

31. Select the MOST EFFECTIVE action (a – e) in response to the situation.

32. Select the LEAST EFFECTIVE action (a – e) in response to the situation.

Situation 17

A senior officer approaches you and asks to talk privately about your supervisor. The senior officer says that they're investigating your supervisor, and they want you to answer some questions about their behavior. You respect your supervisor and don't want anything to happen to them, though you have witnessed some questionable behavior that makes you uneasy.

Possible actions:

a. Tell the senior officer you'd rather not answer their questions about your supervisor.
b. Tell the senior officer you'll comply with their request only if your supervisor is also present.
c. Write a letter to the senior officer explaining the behavior of your supervisor to avoid a face-to-face confrontation.
d. Tell the senior officer that you admire your supervisor, but you've recently noticed some questionable behavior.
e. Tell the senior officer that your supervisor seems just fine, and there's nothing that you can think of regarding their behavior.

33. Select the MOST EFFECTIVE action (a – e) in response to the situation.

34. Select the LEAST EFFECTIVE action (a – e) in response to the situation.

Situation 18

One of your subordinates has been coming to work with what smells like alcohol on his breath. He has come in late only twice, but other Airmen have complained that he occasionally slurs his words or doesn't respond to questions. He's well liked and a good Airman when sober.

Possible actions:

a. Fire him immediately. Coming into work drunk is unacceptable, and he should be used as an example for the rest of the unit.
b. Don't do anything. What the Airman does on his own time isn't your business.
c. Confront him about you and your unit's concerns, and inform him of his options to seek help.
d. Send an email out to the entire base describing the situation.
e. Refer him to the base's counselor, but don't tell him why you've recommended the sessions.

35. Select the MOST EFFECTIVE action (a – e) in response to the situation.

36. Select the LEAST EFFECTIVE action (a – e) in response to the situation.

Situation 19

You just received an email from a subordinate that you consider informal and inappropriate. Your unit has standards for email communications, and you believe that the subordinate came off as disrespectful in the email.

Possible actions:

a. Email the subordinate back, and reprimand them for their informality.
b. Send a mass email to the base as an example of the wrong way to send an email.
c. Invite the subordinate to chat with you, and let them know the proper way to send email communications.
d. Tell your superior officer what happened, and ask for their advice on what to do.
e. Gather all your subordinates in your office, and tell them that learning email communications is really important.

37. Select the MOST EFFECTIVE action (a – e) in response to the situation.

38. Select the LEAST EFFECTIVE action (a – e) in response to the situation.

Situation 20

Recently, you did a personal favor for a fellow officer, even though it was outside your job description. He needed help with editing a proposal that was due the next week, and the officer was dealing with a recent death in his family. Now, two more fellow officers have asked you to edit their proposals as well. You find that you barely have time to complete your own work.

Possible actions:

a. Inform the officers that you're too busy with other work, and editing isn't in your job description.
b. Do the work for them since you agreed to help the first officer.
c. Inform them you can't do the work, and tell their superior officers that they're slacking on their job.
d. Do the work for them, but do it poorly so that they'll never ask you for help again.
e. Let them know that you can't edit their proposals, but find someone else who can.

39. Select the MOST EFFECTIVE action (a – e) in response to the situation.

40. Select the LEAST EFFECTIVE action (a – e) in response to the situation.

Situation 21

You've recently been transferred to a new unit that's understaffed and facing administrative and scheduling problems. You won't be here for long—maybe five or six months—but you're still having trouble understanding how to do your job.

Possible actions:

a. Don't do anything—you won't be here for long, so it's best not to waste their time with extra training.
b. Tell them that you can't do your job without some additional guidance, and wait for them to decide the best course of action before continuing with your work.
c. Watch your coworkers, and mimic what they do.
d. Request help from the administrative team—either a mentor or extra training—so that you feel comfortable doing the assignments given to you.
e. Intentionally try and do the worst possible job, so they'll transfer you back to your old unit.

41. Select the MOST EFFECTIVE action (a – e) in response to the situation.

42. Select the LEAST EFFECTIVE action (a – e) in response to the situation.

Situation 22

You've scheduled a meeting with an important visitor to the base, and you're using one of the best conference rooms available. Every other conference room is booked for this particular time by the managing director. When you arrive at the conference room, you see that another meeting is being set up at the same time. When you ask them what they're doing, they tell you that they also booked the conference room for this time.

Possible actions:

a. Pretend like you didn't know that the room was double-booked when your guest arrives.
b. Tell the other team that they must find another space to hold their meeting since you have an important visitor coming.
c. Contact the managing director, and ask them if there are any alternative meeting spaces.
d. Assess the situation by communicating to the current meeting what's at stake, and listen to their situation as well.
e. Reschedule your meeting with the visitor, and let the other team have the conference room.

43. Select the MOST EFFECTIVE action (a – e) in response to the situation.

44. Select the LEAST EFFECTIVE action (a – e) in response to the situation.

Situation 23

You're being honored at a team meeting for your efficient use of resources. You're expecting to receive a promotion after this meeting, which has already been rumored about in your unit. At the same time, your unit has been struggling with their workload, and you believe that the team is understaffed compared to other comparable teams.

Possible actions:

a. Thank everyone at the meeting, and announce that your team is currently facing some issues.
b. Approach one of your superior officers as soon as you can, and express your concerns to them.
c. Ask your team what you should do, and implement the team's decision.
d. Don't do anything; your team can manage for a little while longer, at least until after you've received the promotion.
e. Let your superior officer know that you'll leave the team if the staffing situation isn't fixed.

45. Select the MOST EFFECTIVE action (a – e) in response to the situation.

46. Select the LEAST EFFECTIVE action (a – e) in response to the situation.

Situation 24

You've been asked to take a look at the Air Force website and give feedback on how to better navigate it as a customer. You don't have much experience or knowledge concerning the technical aspect of websites, but you do have experience in navigating a website as a customer.

Possible actions:

a. Briefly look at the website, and write a few notes about its aesthetics.
b. Plan to give the website good reviews. You're sure it's fine as is, and really, what does it matter if it has a few glitches?
c. Do some research on websites, visit the site as a customer, and disclose your limited knowledge on the topic to your superior officer before giving your feedback.
d. Ask your coworkers to help you review the website and determine what can be changed.
e. Tell your superior officer that you can't look at the website since you don't have the appropriate skills to do so.

47. Select the MOST EFFECTIVE action (a – e) in response to the situation.

48. Select the LEAST EFFECTIVE action (a – e) in response to the situation.

Situation 25

You notice that your coworker has been manipulating numbers in one of the reports you conduct together. The numbers estimate what percentage of Airmen is needed for a combat mission, and you believe the mishap could jeopardize the team's safety.

Possible actions:

a. Don't do anything. Your coworker is experienced at what they do, so you have no right to question them.
b. Go to one of your superior officers, and let them know that the numbers are off.
c. Write an email to the coworker, and tell them that you know what's been happening.
d. Tell them that you refuse to work with them until they've admitted they're manipulating the numbers.
e. Address the situation by telling your coworker that you've noticed the numbers are off, ask for an explanation, and if it doesn't make sense, go to your superior officer.

49. Select the MOST EFFECTIVE action (a – e) in response to the situation.

50. Select the LEAST EFFECTIVE action (a – e) in response to the situation.

Answer Explanations

1. C: (Most Effective): Choice *C* is the most effective response. Reminding the subordinate of their responsibilities and asking them to do their job is the best course of action. It's important to try and solve any issues within your team before going to senior leadership.

2. B: (Least Effective): Choice *B* is the least effective response. Saying nothing to the subordinate is a sign of poor leadership, and a lack of confrontation means that the mistake is likely to reoccur. Choice *E* is the second least effective choice since criticizing and threatening the subordinate doesn't demonstrate appropriate leadership behavior. However, Choice *B* is even less effective than Choice *E* since it ensures the problem will keep happening.

3. D: (Most Effective): Choice *D* is the most effective response. You should help the coworker by splitting the work between both teams in an appropriate manner. The success of the assignment as a whole depends on the success of both teams individually, so the appropriate thing to do would be to offer help.

4. A: (Least Effective): Choice *A* is the least effective response. If you were the one in need of help, you'd hope that a coworker would step up to help. In addition, if one team suffers, the end product suffers, so it's in everyone's best interest to offer help.

5. A: (Most Effective): Choice *A* is the most effective response. If something beyond your control is causing you to perform poorly, you should meet with the senior officer and be honest about what's happening.

6. E: (Least Effective): Choice *E* is the least effective response. You should talk to the senior officer before preparing to move and discussing it with others. Being honest with the senior officer about your plans is the best way to proceed. Choice *D* is the second least effective response. Putting on a front to make others believe something that isn't true is dishonest, but Choice *E* is worse since it's dishonest, and it directly undermines the senior officer.

7. B: (Most Effective): Choice *B* is the most effective response. Creating a presentation, asking permission to present it, and accepting the senior officer's final decision upholds the chain-of-command, while also acting in the unit's best interest.

8. C: (Least Effective): Choice *C* is the least effective response. Implementing the organizational structure against the senior officer's wishes is insubordination. This breaks the chain-of-command, sowing distrust and setting up a conflict between you and the senior officer.

9. E: (Most Effective): Choice *E* is the most effective response. The senior officer is in a better position to decide what's the best course of action when dealing with such sensitive information. Choice *D* is the second most effective answer choice since it's typically better to solve your team's problems before going to senior leadership. However, information contained in the confidential file could threaten your unit's security, so a senior officer should be immediately informed of the mistake.

10. B: (Least Effective): Choice *B* is the least effective response. In this situation, you should be honest, own up to your mistake, and call the senior officer yourself. To do otherwise is to be dishonest.

11. C: (Most Effective): Choice *C* is the most effective response. Weighing the advantages and disadvantages of each side and giving your honest opinion on the subject is important, but it's also critical to avoid becoming involved in their personal conflict.

12. D: (Least Effective): Choice *D* is the least effective response. If a senior officer asks your opinion on a matter, it's your responsibility to answer honestly with complete candor.

13. A: (Most Effective): Choice *A* is the most effective response. The most accountable and honest course of action is to put aside your personal issues, admit your shortcomings, and seek help. The feedback and advice you'll receive will also improve your future work.

14. D: (Least Effective): Choice *D* is the least effective response. You should try to resolve the issue between you and the senior officer before taking it to anyone else. Choice *E* is the second least effective response. Blaming your subordinates will make your unit distrust you, and you're also not taking responsibility for your own mistakes, which will cause people to question your integrity. However, Choice *D* is less effective than Choice *E*, because Choice *D* builds distrust and raises integrity questions, while also breaking the chain-of-command.

15. B: (Most Effective): Choice *B* is the most effective response. Talking with your subordinate face-to-face will open up lines of communication, and it lets your subordinate know that their behavior is problematic.

16 C: (Least Effective): Choice *C* is the least effective response. The behavior of your subordinate is an internal issue that doesn't require the attention of your superior officer. If you do so, your superior officer will doubt if you're capable of fulfilling your duties as an officer. Choice *A* is the second least effective response. Publicly scolding your subordinate could raise questions about your integrity, but Choice *C* is the least effective since it needlessly wastes your superior officer's time.

17. C: (Most Effective): Choice *C* is the most effective response. It's important to have a conversation with your subordinate before doing anything rash because perhaps they don't realize their behavior is negatively impacting the unit.

18. E: (Least Effective): Choice *E* is the least effective response. The situation is affecting the unit as a whole, so it must be addressed in some way.

19. E: (Most Effective): Choice *E* is the most effective response. Meeting with your superior officer and letting them know that you're ready to take on more independence is the best way to express what you need in order to do your job.

20. B: (Least Effective): Choice *B* is the least effective response. Avoiding your superior officer will cause confusion, and it may lead to distrust growing between the two of you.

21. D: (Most Effective): Choice *D* is the most effective response. It's the most honest way to handle this difficult situation. It is always best to try to avoid going behind your partner's back, so telling them first is important, and your superior officer should know if there's a problem that can't be solved without their intervention.

22. E: (Least Effective): Choice *E* is the least effective response. There's already a problem—your partner is still behind on his work despite you doing more than your fair share. This needs to be addressed before it gets any worse, or else the project could be seriously delayed. Choice *A* is the second least effective response. You're already helping your partner, so continuing to take on additional work could hurt the quality of your work. However, Choice *E* is the least effective, because pretending there isn't a problem is dishonest and likely to backfire.

23. E: (Most Effective): Choice *E* is the most effective response. It's the most honest approach. Your superior officer is asking for your honest opinion, so the correct response would be to share your honest opinion.

24. C: (Least Effective): Choice *C* is the least effective response. You know the second candidate isn't as qualified as the first one, so you must say that to your superior officer to avoid lying. The effectiveness of senior leadership depends on their inferior officers' honesty and candor.

25. C: (Most Effective): Choice *C* is the most effective response. It's probably a mistake that you didn't receive your equipment, so it's appropriate to inform your superior officer that you didn't receive the proper resources needed to do your job effectively.

26. E: (Least Effective): Choice *E* is the least effective response. This is an overreaction to what's almost certainly a mistake and needlessly causes a conflict. The simplest solution is to ask your superior officer for the equipment, rather than requesting a transfer to a whole new unit.

27. D: (Most Effective): Choice *D* is the most effective response. Approaching the officer and giving him a chance to explain what's been going on allows you to get the whole picture. The situation might be different than what you've assumed.

28. B: (Least Effective): Choice *B* is the least effective response. If you do nothing, then items will continue to go missing from the supply room.

29. C: (Most Effective): Choice *C* is the most effective response. If they follow your request and delete the email, it rectifies the situation, but you should still be honest with your superior officer since the confidential information could pose a security risk.

30. D: (Least Effective): Choice *D* is the least effective response. This is by far the most dishonest response, and it destroys your integrity. If you're willing to lie about a relatively minor honest mistake, then you can't be trusted as an officer. Choice *B* is the second least effective response. Pretending like the situation never happened is dishonest and ineffective, but fabricating accusations against a subordinate is much worse.

31. A: (Most Effective): Choice *A* is the most effective response. You should always respond to a senior officer's inquiries, but it's important to disclose your lack of experience or knowledge on the topic.

32. E: (Least Effective): Choice *E* is the least effective response. Refusing to answer a senior officer is insubordination, and it could be interpreted as a personal slight against the senior officer.

33. D: (Most Effective): Choice *D* is the most effective response. Total honesty is the best way to handle this uncomfortable situation.

34. E: (Least Effective): Choice *E* is the least effective response. This is not only a lie, but you're now complicit in whatever your supervisor is doing. You always want to be completely upfront with a senior officer, especially if they're conducting an investigation.

35. C: (Most Effective): Choice *C* is the most effective response. If your subordinate is coming into work drunk, your subordinate probably has an alcohol problem and needs to seek help, such as counseling or treatment.

36. B: (Least Effective): Choice *B* is the least effective response. If a subordinate is drunk at work, then it's your responsibility to take action. The Airman could pose a risk to himself or other Airmen if the problem is left unaddressed.

37. C: (Most Effective): Choice *C* is the most effective response. It clearly notifies the subordinate that there's a problem and gives them a solution to fix it.

38. B: (Least Effective): Choice *B* is the least effective response. This would embarrass the subordinate, and it may cause others to distrust you in the future.

39. A: (Most Effective): Choice *A* is the most effective response. You should let the officers know that you can't take on an extracurricular assignment when you're struggling with your current workload.

40. D: (Least Effective): Choice *D* is the least effective response. If you agree to do work for someone else, you should do it to the best of your ability. Intentionally sabotaging other people's work is dishonest, and in addition, the work's poor quality would harm your reputation.

41. D: (Most Effective): Choice *D* is the most effective response. If you can't complete your work due to a lack of knowledge, you should ask for help. Anything less would waste your time and hurt your unit.

42. E: (Least Effective): Choice *E* is the least effective response. Intentionally producing unsatisfactory work reflects poorly on your character and abilities.

43. D: (Most Effective): Choice *D* is the most effective response. Communicating with the other team could lead to a workable solution for all parties involved. Choice *C* is the second most effective response, but the matter should be discussed with the team currently in the room before approaching the managing director.

44. A: (Least Effective): Choice *A* is the least effective response. This is dishonest, and it'll make you look incompetent once the visitor arrives. In addition, ignoring the problem wastes your important visitor's time.

45. B: (Most Effective): Choice *B* is the most effective response. It's best to be honest about what's happening right away, even though you're being honored for your efficient leadership approach. Waiting will cause more problems in the end.

46. D: (Least Effective): Choice *D* is the least effective response. Not doing anything will not fix the problem, and it could make your team members distrust your leadership.

47. C: (Most Effective): Choice *C* is the most effective response. The preliminary research shows some initiative, and disclosing your lack of knowledge retains your honesty without refusing to do your assigned work.

48. B: (Least Effective): Choice *B* is the least effective response. Your superior officer clearly believes it does matter whether or not the website has glitches, and they're relying on your honest assessment to improve the website. Choice *E* is the second least effective response, because you shouldn't outright refuse your superior officer's orders. However, Choice *B* is the least effective response, because the lying compromises your integrity and undermines the Air Force's recruitment on the website.

49. E: (Most Effective): Choice *E* is the most effective response. You should approach your coworker before bringing the matter to a senior officer. Maybe you've misread the report, or maybe there's a good explanation for what's been happening. If the numbers still don't make sense after the conversation, you should then talk to a senior officer.

50. A: (Least Effective): Choice *A* is the least effective response. If the data is incorrect, then team members' lives could be in danger. It's your duty to either find out what's happening or bring the matter to the attention of someone who can do so.

Self-Description Inventory

The self-description inventory provided in the AFOQT has no impact on your numerical score. Instead, these questions help discover your specific personality traits. With this in mind, you yield more accurate results if you select answers that fit your first instinct. Don't overthink this. These results will help assess the best career fit for you in the Air Force by comparing your results to that of other Air Force personnel.

Physical Science

Chemistry

Scientific Notation, the Metric System, and Temperature Scales

Scientific Notation

Scientific notation is a system used to represent numbers that are very large or very small. Sometimes, numbers are way too big or small to be written out with multiple zeros behind them or in decimal form, so scientific notation is used as a way to express these numbers in a simpler way.

Scientific notation takes the decimal notation and turns it into scientific notation, like the table below:

Decimal Notation	Scientific Notation
5	5×10^0
500	5×10^2
10,000,000	1×10^7
8,000,000,000	8×10^9
-55,000	-5.5×10^4
.00001	10^{-5}

In scientific notation, the decimal is placed after the first digit and all the remaining numbers are dropped. For example, 5 becomes "5.0×10^0." This equation is raised to the zero power because there are no zeros behind the number "5." Always put the decimal after the first number. Let's say we have the number 125,000. We would write this using scientific notation as follows: 1.25×10^5, because to move the decimal from behind "1" to behind "125,000" takes five counts, so we put the exponent "5" behind the "10." As you can see in the table above, the number ".00001" is too cumbersome to be written out each time for an equation, so we would want to say that it is "10^{-5}." If we count from the place behind the decimal point to the number "1," we see that we go backwards 5 places. Thus, the "-5" in the scientific notation form represents 5 places to the right of the decimal.

Converting Within and Between Standard and Metric Systems

Recall that the metric system has base units of meter for length, kilogram for mass, and liter for liquid volume. This system expands to three places above the base unit and three places below. These places correspond with prefixes with a base of 10. The following table shows the conversions:

kilo-	hecto-	deka-	base	deci-	centi-	milli-
1,000 times the base	100 times the base	10 times the base		1/10 times the base	1/100 times the base	1/1000 times the base

To convert between units within the metric system, values with a base ten can be multiplied. The decimal can also be moved in the direction of the new unit by the same number of zeros on the number. For example, 3 meters is equivalent to .003 kilometers. The decimal moved three places (the same number of zeros for kilo-) to the left (the same direction from base to kilo-). Three meters is also equivalent to 3,000 millimeters. The decimal is moved three places to the right because the prefix milli- is three places to the right of the base unit.

The English Standard system used in the United States has a base unit of foot for length, pound for weight, and gallon for liquid volume. These conversions aren't as easy as the metric system because they aren't a base ten model. The following table shows the conversions within this system:

Length	Weight	Capacity
1 foot (ft) = 12 inches (in) 1 yard (yd) = 3 feet 1 mile (mi) = 5280 feet 1 mile = 1760 yards	1 pound (lb) = 16 ounces (oz) 1 ton = 2000 pounds	1 tablespoon (tbsp) = 3 teaspoons (tsp) 1 cup (c) = 16 tablespoons 1 cup = 8 fluid ounces (oz) 1 pint (pt) = 2 cups 1 quart (qt) = 2 pints 1 gallon (gal) = 4 quarts

When converting within the English Standard system, most calculations include a conversion to the base unit and then another to the desired unit. For example, take the following problem: 3 $quarts =$ ___$cups$. There is no straight conversion from quarts to cups, so the first conversion is from quarts to pints. There are 2 pints in 1 quart, so there are 6 pints in 3 quarts. This conversion can be solved as a proportion:

$$\frac{3 \; qt}{x} = \frac{1 \; qt}{2 \; pints}$$

It can also be observed as a ratio 2:1, expanded to 6:3. Then the 6 pints must be converted to cups. The ratio of pints to cups is 1:2, so the expanded ratio is 6:12. For 6 pints, the measurement is 12 cups. This problem can also be set up as one set of fractions to cancel out units. It begins with the given information and cancels out matching units on top and bottom to yield the answer. Consider the following expression:

$$\frac{3 \; quarts}{1} \times \frac{2 \; pints}{1 \; quart} \times \frac{2 \; cups}{1 \; pint}$$

It's set up so that units on the top and bottom cancel each other out:

$$\frac{3 \; \cancel{quarts}}{1} \times \frac{2 \; \cancel{pints}}{1 \; \cancel{quart}} \times \frac{2 \; cups}{1 \; \cancel{pint}}$$

The numbers can be calculated as 3 × 3 × 2 on the top and 1 on the bottom. It still yields an answer of 12 cups.

This process of setting up fractions and canceling out matching units can be used to convert between standard and metric systems. A few common equivalent conversions are 2.54 cm = 1 inch, 3.28 feet = 1 meter, and 2.205 pounds = 1 kilogram. Writing these as fractions allows them to be used in conversions. For the fill-in-the-blank problem 5 meters = ___ feet, an expression using conversions starts with the expression

$$\frac{5 \; meters}{1} \times \frac{3.28 \; feet}{1 \; meter}$$

where the units of meters will cancel each other out, and the final unit is feet. Calculating the numbers yields 16.4 feet. This problem only required two fractions. Others may require longer expressions, but the underlying rule stays the same. When there's a unit on the top of the fraction that's the same as the unit on the bottom, then they cancel each other out. Using this logic and the conversions given above, many units can be converted between and within the different systems.

Temperature Scales

Science utilizes three primary temperature scales. The temperature scale most often used in the United States is the Fahrenheit (F) scale. The Fahrenheit scale uses key markers based on the measurements of the freezing (32 °F) and boiling (212 °F) points of water. In the United States, when taking a person's temperature with a thermometer, the Fahrenheit scale is used to represent this information. The human body registers an average temperature of 98.6 °F.

Another temperature scale commonly used in science is the Celsius (C) scale (also called *centigrade* because the overall scale is divided into one hundred parts). The Celsius scale marks the temperature for water freezing at 0 °C and boiling at 100 °C. The average temperature of the human body registers at 37 °C. Most countries in the world use the Celsius scale for everyday temperature measurements.

For scientists to easily communicate information regarding temperature, an overall standard temperature scale was agreed upon. This scale is the Kelvin (K) scale. Named for Lord Kelvin, who conducted research in thermodynamics, the Kelvin scale contains the largest range of temperatures to facilitate any possible readings.

The Kelvin scale is the accepted measurement by the International System of Units (from the French *Système international d'unités*), or SI, for temperature. The Kelvin scale is employed in thermodynamics, and its reading for 0 is the basis for absolute zero. This scale is rarely used for measuring temperatures in the medical field.

The conversions between the temperature scales are as follows:

Degrees Fahrenheit to Degrees Celsius:

$$ {}^0C = \frac{5}{9}({}^0F - 32) $$

Degrees Celsius to Degrees Fahrenheit:

$$ {}^0F = \frac{9}{5}({}^\circ C) + 32 $$

Degrees Celsius to Kelvin:

$$ K = {}^0C + 273.15 $$

For example, if a patient has a temperature of 38 °C, what would this be on the Fahrenheit scale?

Solution:

First, select the correct conversion equation from the list above.

$$ {}^0F = \frac{9}{5}({}^\circ C) + 32 $$

Next, plug in the known value for °C, 38.

$$ {}^0F = \frac{9}{5}(38) + 32 $$

Finally, calculate the desired value for °F.

$$°F = \frac{9}{5}(38) + 32$$

$$°F = 100.4°F$$

For example, what would the temperature 52 °C be on the Kelvin scale?

First, select the correct conversion equation from the list above.

$$K = °C + 273.15$$

Next, plug in the known value for °C, 52.

$$K = 52 + 273.15$$

Finally, calculate the desired value for K.

$$K = 325.15 \ K$$

Atomic Structure and the Periodic Table

Today's primary model of the atom was proposed by scientist Niels Bohr. Bohr's atomic model consists of a nucleus, or core, which is made up of positively charged protons and neutrally charged neutrons. Neutrons are theorized to be in the nucleus with the protons to provide "balance" and stability to the protons at the center of the atom. More than 99 percent of the mass of an atom is found in the nucleus. Orbitals surrounding the nucleus contain negatively charged particles called **electrons**. Since the entire structure of an atom is too small to be seen with the unaided eye, an electron microscope is required for detection. Even with such magnification, the actual particles of the atom are not visible.

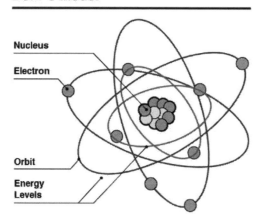

Bohr's Model

Anything that takes up space and has mass is considered *matter*. Matter is composed of atoms. An **atom** has an atomic number that is determined by the number of protons within the nucleus. **Properties**, which are observable characteristics of a substance, can be used to describe different substances. The **physical properties** of a substance can be observed without altering the identity or chemical composition of the substance. Density, solubility, malleability, odor, and luster are examples of physical properties. **Chemical properties,** on the other hand, can only be observed through a change in the chemical composition of the substance via a chemical reaction. Flammability, oxidative state, and reactivity with an acid are examples of chemical properties.

Some substances are made up of atoms, all with the same atomic number. Such a substance is called an **element.** Using their atomic numbers, elements are organized and grouped by similar properties in a chart called the **Periodic Table.**

The sum of the total number of protons in an atom and the total number of neutrons in the atom provides the atom's **mass number**. Most atoms have a nucleus that is electronically neutral, and all atoms of one type have the same atomic number. There are some atoms of the same type that have a different mass number. The variation in the mass number is due to an imbalance of neutrons within the nucleus of the atoms. If atoms have this variance in neutrons, they are called **isotopes.** It is the different number of neutrons that gives such atoms a different mass number. For isotopes, the atomic number (determined by the number of protons) is the same, but the mass number (determined by adding the protons and neutrons) is different.

This is a result of there being a different number of neutrons.

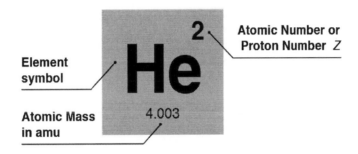

A concise method of arranging elements by atomic number, similar characteristics, and electron configurations in a tabular format was necessary to represent elements. This was originally organized by scientist Dmitri Mendeleev using the Periodic Table. The vertical lines on the Periodic Table are called **groups** and are sorted by similar chemical properties/characteristics, such as appearance and reactivity. This is observed in the shiny texture of metals, the softness of post-transition metals, and the high melting points of alkali earth metals. The horizontal lines on the Periodic Table are called **periods** and are arranged by electron valance configurations.

The Periodic Table of the Elements

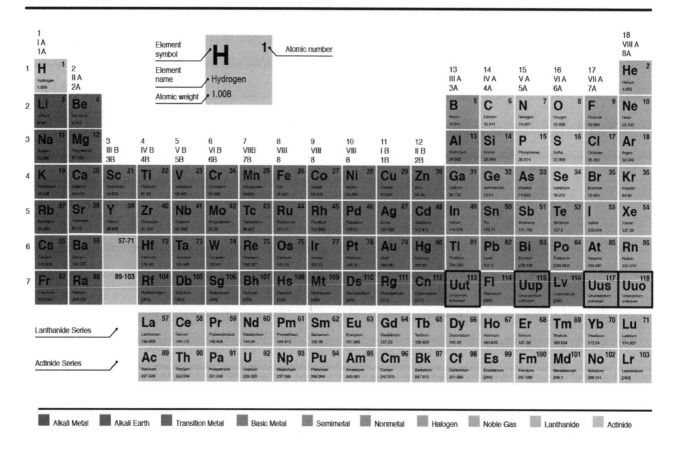

Elements are set by ascending atomic number, from left to right. The number of protons contained within the nucleus of the atom is represented by the atomic number. For example, hydrogen has one proton in its nucleus, so it has an atomic number of 1.

Since isotopes can have different masses within the same type of element, the atomic mass of an element is the average mass of all the naturally occurring atoms of that given element. **Atomic mass** is calculated by finding the relative abundance of isotopes that might be used in chemistry, or by adding the number of protons and neutrons of an atom together. For example, the atomic number of chlorine is 35 on the Periodic Table. However, the atomic mass of chlorine is 35.5 atomic mass units (amu). This discrepancy occurs because a large number of chlorine isotopes (meaning, instead of 35 neutrons, a chlorine nucleus might have 36 neutrons) exist in nature. The average of all the atomic masses turns out to be 35.5 amu, which is slightly higher than chlorine's listed number on the Periodic Table. Carbon has an atomic number of 12, but its atomic mass is 12.01 amu because there are not as many naturally occurring isotopes to raise the average number, as observed with chlorine.

Chemical Reactions

A **chemical reaction** is a process that involves a change in the molecular arrangement of a substance. Generally, one set of chemical substances, called the **reactants**, is rearranged into a different set of chemical substances, called the **products**, by the breaking and re-forming of bonds between atoms. In a chemical reaction, it is important to realize that no new atoms or molecules are introduced. The products are formed solely from the atoms and molecules that are present in the reactants. These can involve a change in state of matter as well. Making glass, burning fuel, and brewing beer are all examples of chemical reactions.

Generally, chemical reactions are thought to involve changes in positions of electrons with the breaking and re-forming of chemical bonds, without changes to the nucleus of the atoms. The four main types of chemical reactions are combination, decomposition, combustion, and oxidation/reduction reactions.

Combination
In combination reactions, two or more reactants are combined to form one more complex, larger product. The bonds of the reactants are broken, the elements are arranged, and then new bonds are formed between all of the elements to form the product. It can be written as

$$A + B \rightarrow C$$

where A and B are the reactants and C is the product. An example of a combination reaction is the creation of iron(II) sulfide from iron and sulfur, which is written as $8Fe + S_8 \rightarrow 8FeS$.

Decomposition
Decomposition reactions are almost the opposite of combination reactions. They occur when one substance is broken down into two or more products. The bonds of the first substance are broken, the elements are rearranged, and then the elements are bonded together in new configurations to make two or more molecules. These reactions can be written as

$$C \rightarrow B + A$$

where C is the reactant and A and B are the products. An example of a decomposition reaction is the electrolysis of water to make oxygen and hydrogen gas, which is written as

$$2H_2O \rightarrow 2H_2 + O_2.$$

Combustion
Combustion reactions are a specific type of chemical reaction that involves oxygen gas as a reactant. This mostly involves the burning of a substance. The combustion of hexane in air is one example of a combustion reaction. The hexane gas combines with oxygen in the air to form carbon dioxide and water. The reaction can be written as

$$2C_6H_{14} + 17O_2 \rightarrow 12CO_2 + 14H_2O.$$

Oxidation and Reduction
Oxidation/reduction (redox or half) reactions involve the oxidation of one species and the reduction of the other species in the reactants of a chemical equation. This can be seen through three main types of transfers.

The first type is through the transfer of oxygen. The reactant gaining an oxygen is the oxidizing agent, and the reactant losing an oxygen is the reduction agent.

For example, the oxidation of magnesium is as follows:

$$2\ Mg\,(s) + O_2\,(g) \rightarrow\ 2\ MgO\,(s)$$

The second type is through the transfer of hydrogen. The reactant losing the hydrogen is the oxidizing agent, and the other reactant is the reduction agent.

For example, the redox of ammonia and bromine results in nitrogen and hydrogen bromide due to bromine gaining a hydrogen as follows:

$$2\ NH_3 + 3\ Br_2 \rightarrow\ N_2 + 6\ HBr$$

The third type is through the loss of electrons from one species, known as the **oxidation agent,** and the gain of electrons to the other species, known as the **reduction agent.** For a reactant to become "oxidized," it must give up an electron.

For example, the redox of copper and silver is as follows:

$$Cu\textit{(s)} + 2\ Ag^+\textit{(aq)} \rightarrow\ Cu^{2+}\textit{(aq)} + 2\ Ag\textit{(s)}$$

It is also important to note that the oxidation numbers can change in a redox reaction due to the transfer of oxygen atoms. Standard rules for finding the oxidation numbers for a compound are listed below:

1. The oxidation number of a free element is always 0.

2. The oxidation number of a monatomic ion equals the charge of the ion.

3. The oxidation number of H is +1, but it is –1 when combined with less electronegative elements.

4. The oxidation number of O in compounds is usually –2, but it is –1 in peroxides.

5. The oxidation number of a Group 1 element in a compound is +1.

6. The oxidation number of a Group 2 element in a compound is +2.

7. The oxidation number of a Group 17 element in a binary compound is –1.

8. The sum of the oxidation numbers of all the atoms in a neutral compound is 0.

9. The sum of the oxidation numbers in a polyatomic ion is equal to the charge of the ion.

These rules can be applied to determine the oxidation number of an unknown component of a compound.

For example, what is the oxidation number of Cr in $CrCl_3$?

From rule 7, the oxidation number of Cl is given as –1. Since there are 3 chlorines in this compound, that would equal 3 × –1 for a result of –3. According to rule 8, the total oxidation number of Cr must balance the total oxidation number of Cl, so Cr must have a total oxidation number equaling +3 ($-3 + +3 = 0$). There is only 1 Cr, so the oxidation number would be multiplied by 1, or the same as the total of +3, written as follows:

+3 -1
$$CrCl_3$$
+3 -3

Chemical Equations

Chemical equations describe how the molecules are changed when the chemical reaction occurs. For example, the chemical equation of the hexane combustion reaction is $2C_6H_{14} + 17O_2 \rightarrow 12CO_2 + 14H_2O$. The "+" sign on the left side of the equation indicates that those molecules are reacting with each other, and the arrow, "$\rightarrow$," in the middle of the equation indicates that the reactants are producing something else. The coefficient before a molecule indicates the quantity of that specific molecule that is present for the reaction. The subscript next to an element indicates the quantity of that element in each molecule. In order for the chemical equation to be balanced, the quantity of each element on both sides of the equation should be equal. For example, in the hexane equation above, there are twelve carbon elements, twenty-eight hydrogen elements, and thirty-four oxygen elements on each side of the equation. Even though they are part of different molecules on each side, the overall quantity is the same. The state of matter of the reactants and products can also be included in a chemical equation and would be written in parentheses next to each element as follows: gas (g), liquid (l), solid (s), and dissolved in water, or aqueous (aq).

Reaction Rates, Equilibrium, and Reversibility

The rate of a chemical reaction can be increased by adding a catalyst to the reaction. **Catalysts** are substances that lower the activation energy required to go from the reactants to the products of the reaction but are not consumed in the process. The **activation energy** of a reaction is the minimum amount of energy that is required to make the reaction move forward and change the reactants into the products. When catalysts are present, less energy is required to complete the reaction. For example, hydrogen peroxide will eventually decompose into two water molecules and one oxygen molecule. If potassium permanganate is added to the reaction, the decomposition happens at a much faster rate. Similarly, increasing the temperature or pressure in the environment of the reaction can increase the rate of the reaction. Higher temperatures increase the number of high-energy collisions that lead to the products. The same happens when increasing pressure for gaseous reactants, but not with solid or liquid reactants. Increasing the concentration of the reactants or the available surface area over which they can react also increases the rate of the reaction.

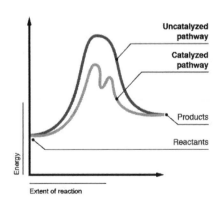

Many reactions are **reversible,** which means that the products can revert back to the reactants under certain conditions. In such reactions, the arrow will be double-headed, indicating that the reaction can proceed in either direction. Reactions reach a state of **equilibrium** when the net concentration of the reactants and products are not changing. This does not mean that the reactions have stopped occurring, but that there is no overall change in either direction (forming more product from the reactants or the product undergoing the reverse reaction to re-form the reactants). Some number of reactions may be going on in both directions, but they cancel each other out so that there is no net change in concentrations on either side of the reaction arrow.

Solutions and Solubility

A **solution** is a homogenous mixture of two or more substances. Unlike heterogenous mixtures, in solutions, the **solute**, which is a substance that can be dissolved, is uniformly distributed throughout the *solvent*, which is the substance in which the solvent dissolves. For example, when 10 grams of table salt (NaCl) is added to a 100 mL of room temperature water and then stirred until all of the salt (the solute) has dissolved in the water (the solvent), a solution is formed. The dissolved salt, in the form of Na^+ and Cl^- ions, will be evenly distributed throughout the water. In this case, the solution is **diluted** because only a small amount of solute was dissolved in a comparatively large volume of solvent. The salt water solution would be said to be **concentrated** if a large amount of salt was added, stirred, and dissolved into the water, 30 grams, for example. When more solute is added to the solvent but even after stirring, some settles on the bottom without dissolving, the solution is **saturated**. For example, in 100 mL of room temperature water, about 35 grams of table salt can dissolve before the solution is saturated. Beyond this point—called the saturation point—any additional salt added will not readily dissolve. Sometimes, it is possible to temporarily dissolve excessive solute in the solvent, which creates a **supersaturated** solution. However, as soon as this solution is disturbed, the process of **crystallization** will begin and a solid will begin precipitating out of the solution.

It is often necessary, for example when working with chemicals or mixing acids and bases, to quantitatively determine the concentration of a solution, which is a more precise measure than using qualitative terms like diluted, concentrated, saturated, and supersaturated. The **molarity**, c, of a solution is a measure of its concentration; specifically, it is the number of moles of solute (represented by n in the formula) per liter of solution (V). Therefore, the following is the formula for calculating the molarity of a solution:

$$c = \frac{n}{V}$$

It is important to remember that the volume in the denominator of the equation above is in liters of solution, not solvent. Adding solute increases the volume of the entire solution, so the molarity formula accounts for this volumetric increase.

Factors Affecting Solubility
Solubility refers to a solvent's ability to dissolve a solute. Certain factors can affect solubility, such as temperature and pressure. Depending on the state of matter of the molecules in the solution, these factors may increase or decrease solubility. When most people think of solutions, they imagine a solid or liquid solute dissolved in a liquid solvent, like water. However, solutions can be composed on molecules in the other states of matter as well. The following are examples of solutions involving combinations of solids, liquids, and gasses:

- Dry air is solution of gasses, mainly nitrogen and oxygen, with carbon dioxide and others as well

- Vinegar is a solution of liquid acetic acid and liquid water

- Brass is a solution of solid copper and solid zinc

- Amalgams can be a solution of liquid mercury with a solid metal like gold

- Hummingbirds like sweet water, a solution of liquid water and solid sucrose (table sugar)

Increases in temperature, tend to increase a solute's solubility, except in the case of gasses wherein higher temperature reduce the solubility of gaseous solutions and gasses dissolved in water. However, the solubility of gasses in organic solvents increases with increases in temperature. Pressure is directly related to the solubility of gasses, but not of liquids or solids. Agitating, or stirring, a solution does not affect solubility, but it does increase the rate at which the solute dissolves.

The electronegativities, or **polarities**, of the solvent and solute determine whether the solute will readily dissolve in the solvent. The key to the potential for a solution to form and the solute to dissolve in the solvent is "like dissolves like." If the solute is polar, it will dissolve in a polar solvent. If the solute is non-polar, it will dissolve in a non-polar solvent. Water is an example of a polar solvent, while benzene is an example of a non-polar solvent.

Stoichiometry

Stoichiometry uses proportions based on the principles of the conservation of mass and the conservation of energy. It deals with first balancing the chemical (and sometimes physical) changes in a reaction and then finding the ratios of the reactants used and what products resulted. Just as there are different types of reactions, there are different types of stoichiometry problems. Different reactions can involve mass, volume, or moles in varying combinations. The steps to solve a stoichiometry problem are to first, balance the equation; next, find the number of total products; and finally, calculate the desired information regarding molar mass, percent yield, etc.

The **molar mass** of any substance is the measure of the mass of one mole of that substance. For pure elements, the molar mass is also referred to as the **atomic mass unit (amu)** for that substance. In compounds, this is calculated by adding the molar mas of each substance in the compound. For example, the molar mass of carbon can be found on the Periodic Table as 12.01 g/mol, while finding the molar mass of water (H_2O) requires a bit of calculation.

$$
\begin{array}{lll}
\text{the molar mass of hydrogen} = & 1.01 \times 2 & = 2.02 \\
+ \text{ the molar mass of oxygen} = & 16.0 & = \underline{+16.00} \\
& & 18.02 \text{ g/mol}
\end{array}
$$

To determine the **percent composition** of a compound, the individual molar masses of each component need to be divided by the total molar mass of the compound and then multiplied by 100. For example, to find the percent composition of carbon dioxide (CO_2) first requires the calculation of the molar mass of CO_2.

$$
\begin{array}{lll}
\text{the molar mass of carbon} = & 12.01 & = 12.01 \\
+ \text{ the molar mass of oxygen} = & 16.0 \times 2 & = \underline{+32.00} \\
& & 44.01 \text{ g/mol}
\end{array}
$$

Next, take each individual mass, divide by the total mass of the compound, and then multiply by 100 to get the percent composition of each component.

$$\frac{12.01 \ g/mol}{44.01 \ g/mol} = 0.2729 \times 100 = 27.29\% \ carbon$$

$$\frac{32.00 \ g/mol}{44.01 \ g/mol} = 0.7271 \times 100 = 72.71\% \ oxygen$$

A quick check in the addition of the percentages should always total 100%.

If an example provides the basis for an equation, the equation would first need to be balanced to calculate any proportions. For example, if 15 g of C_2H_6 reacts with 64 g of O_2, how many grams of CO_2 will be formed?

First, write the chemical equation:

$$C_2H_6 \ + \ O_2 \ \rightarrow \ CO_2 \ + \ H_2O$$

Next, balance the equation:

$$2\,C_2H_6 \ + \ 7\,O_2 \ \rightarrow \ 4\,CO_2 \ + \ 6\,H_2O$$

Then, calculate the desired amount based on the beginning information of 15 g of C2H6:

$$15 \ g \ C_2H_6 \ \times \ \frac{1 \ mole \ C_2H_6}{30 \ g \ C_2H_6} \ \times \ \frac{4 \ moles \ CO_2}{2 \ moles \ C_2H_6} \ \times \ \frac{44 \ g \ CO_2}{1 \ mole \ CO_2} \ = \ 44 \ g \ CO_2$$

To check that this would be the smaller amount (or how much until one of the reactants is used up, thus ending the reaction), the calculation would need to be done for the 64 g of O_2:

$$64 \ g \ O_2 \ \times \ \frac{1 \ mole \ O_2}{32 \ g \ O_2} \ \times \ \frac{4 \ moles \ CO_2}{7 \ moles \ O_2} \ \times \ \frac{44 \ g \ CO_2}{1 \ mole \ CO_2} \ = \ 50.5 \ g \ CO_2$$

The yield from the C_2H_6 is smaller, so it would be used up first, ending the reaction. This calculation would determine the maximum amount of CO_2 that could possibly be produced.

Acids and Bases

If something has a sour taste, it is considered acidic, and if something has a bitter taste, it is considered basic. Acids and bases are generally identified by the reaction they have when combined with water. An **acid** will increase the concentration of hydrogen ions (H^+) in water, and a **base** will increase the concentration of hydroxide ions (OH^-). Other methods of identification with various indicators have been designed over the years.

To better categorize the varying strength levels of acids and bases, the pH scale is employed. The **pH scale** is a logarithmic (base 10) grading applied to acids and bases according to their strength. The pH scale contains values from 0 through 14 and uses 7 as neutral. If a solution registers below a 7 on the pH scale, it is considered an acid. If a solution registers higher than a 7, it is considered a base. To perform a quick test on a solution, litmus paper can be used. A base will turn red litmus paper blue, and an acid will turn blue litmus paper red. To gauge the strength of an acid or base, a test using phenolphthalein can be administered. An acid will turn red phenolphthalein to colorless, and a base will turn colorless phenolphthalein to pink. As demonstrated with these types of tests, acids and bases neutralize each other. When acids and bases react with one another, they produce salts (also called **ionic substances**).

Acids and bases have varying strengths. For example, if an acid completely dissolves in water and ionizes, forming an H^+ and an anion, it is considered a strong acid. There are only a few common strong acids, including sulfuric (H_2SO_4), hydrochloric (HCl), nitric (HNO_3), hydrobromic (HBr), hydroiodic (HI), and perchloric ($HClO_4$). Other types of acids are considered weak.

An easy way to tell if something is an acid is by looking for the leading "H" in the chemical formula.

A base is considered strong if it completely dissociates into the cation of OH^-, including sodium hydroxide (NaOH), potassium hydroxide (KOH), lithium hydroxide (LiOH), cesium hydroxide (CsOH), rubidium hydroxide (RbOH), barium hydroxide ($Ba(OH)_2$), calcium hydroxide ($Ca(OH)_2$), and strontium hydroxide ($Sr(OH)_2$). Just as with acids, other types of bases are considered weak. An easy way to tell if something is a base is by looking for the "OH" ending on the chemical formula.

In pure water, autoionization occurs when a water molecule (H_2O) loses the nucleus of one of the two hydrogen atoms to become a hydroxide ion (OH^-). The nucleus then pairs with another water molecule to form hydronium (H_3O^+). This autoionization process shows that water is **amphoteric**, which means it can react as an acid or as a base.

Pure water is considered neutral, but the presence of any impurities can throw off this neutral balance, causing the water to be slightly acidic or basic. This can include the exposure of water to air, which can introduce carbon dioxide molecules to form carbonic acid (H_2CO_3), thus making the water slightly acidic. Any variation from the middle of the pH scale (7) indicates a non-neutral substance.

Nuclear Chemistry

Nuclear chemistry (also referred to as **nuclear physics**) deals with interactions within the nuclei of atoms. This differs from typical chemical reactions, which involve interactions with the electrons of atoms. If the nucleus of an atom is unstable, it emits radiation as it releases energy resulting from changes in the nucleus. This instability often occurs in isotopes of an element. An **isotope** is formed when the nucleus of an atom has the same number of protons but a different number of neutrons. This difference in mass causes a heavy, unstable condition in the nucleus.

According to quantum theory, there is no way to precisely predict when an atom will decay, but the decay of a collection of atoms in a substance can be predicted by their collective half-life. Half-life is used to calculate the time it takes for have nuclei in atoms of a radioactive substance to have undergone radioactive decay (in which they emit particles and energy).

There are three primary types of nuclear decay occurring in an atom with an unstable nucleus: alpha, beta, and gamma.

In **alpha decay**, an atom will emit two protons and two neutrons from its nucleus. This emission is in the form of a "bundle" and is called an **alpha particle**. Occurring mainly in larger, heavier atoms, alpha decay causes the atom's proton count to drop by two, thus resulting in the creation of a new element. Alpha radiation is extremely weak and can be blocked by something as thin as a piece of paper.

When the neutron of an atom emits an electron and causes the electron count to increase, the proton count of the nucleus is also increased. This action creates a new element and emits **beta radiation** in the process. Beta radiation is slightly more dangerous than alpha radiation, but it can be blocked by heavy materials such as aluminum or even wood.

The most dangerous type of radiation results from gamma decay. **Gamma decay** does not alter the mass or the charge of an atom. Gamma radiation is emitted along with an alpha or beta particle. It is extremely dangerous and can only be blocked by lead.For example, iodine is a stable element, often used in nuclear medicine. Iodine's atomic number is 53, and its atomic mass is 127 amu. When an isotope of iodine, with atomic number 53 and atomic mass of 131 amu (due to an excess of four neutrons), is used in the human body, it can be seen in the thyroid as a radioactive tracer.

When naturally-occurring radioactive elements decay, they follow a radioactive decay series, which starts as one element that decays into a second element, which then decays into a third element, and so on until a stable element is finally reached. There are three naturally-occurring radioactive decay series. Each of these series begins with either uranium-235, uranium-238, or thorium. For example, uranium-238 decays into astatine, bismuth, lead, polonium, protactinium, radium, radon, thallium, and thorium before finally becoming a more stable element. An artificial radioactive series begins with the artificially-made element of neptunium and decays into elements such as polonium and americium on its way to becoming more stable.

When naturally occurring radioactive elements decay, they follow a radioactive decay series, which starts as one element, decays into a second element, decays into a third element, and so on until a stable element is finally reached. There are three naturally occurring radioactive decay series. Each of these series begins with either uranium-235, uranium-238, or thorium. For example, uranium-238 decays into astatine, bismuth, lead, polonium, protactinium, radium, radon, thallium, and thorium before finally becoming a more stable element. An artificial radioactive series begins with the artificially made element of neptunium and decays into elements such as polonium and americium on its way to becoming more stable.

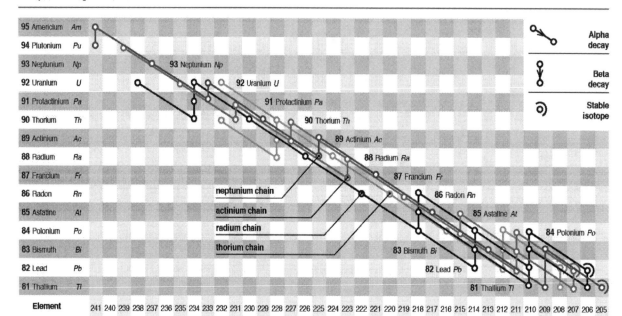

Isotopes can be created through separation and synthesis. There are approximately two hundred radioisotopes; most of them are artificially created. Bombarding the nucleus with nuclear particles or nuclei changes the nucleus of an element. This method is called **transmutation** and has resulted in the formation of new, artificial elements.

The application of half-life knowledge can be seen through the carbon cycle used in carbon dating. Some elements have been estimated to be as old as our universe, and studying the position of something such as carbon in its half-life decay cycle can provide an accurate age of a substance containing carbon. Carbon-14 is utilized for this specific purpose.

The understanding of nuclear chemistry is important for scientists to design and utilize radioactive drugs for treatments and diagnostic techniques. If an isotope is radioactive, it can easily be detected as a contrast agent in the human body, enabling medical professionals to view and diagnose issues that could not be observed without contrast or through physical exams or x-rays. When radioactive tracers are introduced, they can aid in the detection of how systems work. Since the amount of radiation used is small, it can help create a map of its path through a system without causing any disruption to that system.

The radioactive materials can also provide the necessary backdrop or contrast for reading emissions from samples by spectroscopes.

Exposure to radiation has varying effects on the human body. Medical treatments can utilize short exposures to radiation, and damage can be localized to specific, targeted areas, as in treatments for cancer. Extended exposure can cause chromosome damage through breaking the chemical bonds of DNA. Prolonged exposure to radiation can even cause cancer by the mutation and killing of cells and through diminishing the body's ability to produce cells and heal. Overexposure to radiation can result in death.

Relationships Among Events, Objects, and Procedures

When we determine relationships among events, objects, and procedures, we are better able to understand the world around us and make predictions based on that understanding. With regards to relationships among events and procedures, we will look at cause and effect.

Cause

The **cause** of a particular event is the thing that brings it about. A causal relationship may be partly or wholly responsible for its effect, but sometimes it's difficult to tell whether one event is the sole cause of another event. For example, lung cancer can be caused by smoking cigarettes. However, sometimes lung cancer develops even though someone does not smoke, and that tells us that there may be other factors involved in lung cancer besides smoking. It's also easy to make mistakes when considering causation. One common mistake is mistaking correlation for causation. For example, say that in the year 2008 a volcano erupted in California, and in that same year, the number of infant deaths increased by ten percent. If we automatically assume, without looking at the data, that the erupting volcano *caused* the infant deaths, then we are mistaking correlation for causation. The two events might have happened at the same time, but that does not necessarily mean that one event caused the other. Relationships between events are never absolute; there are a myriad of factors that can be traced back to their foundations, so we must be thorough with our evidence in proving causation.

Effect

An **effect** is the result of something that occurs. For example, the Nelsons have a family dog. Every time the dog hears thunder, the dog whines. In this scenario, the thunder is the cause, and the dog's whining is the effect. Sometimes a cause will produce multiple effects. Let's say we are doing an experiment to see what the effects of exercise are in a group of people who are not usually active. After about four weeks, the group experienced weight loss, rise in confidence, and higher energy. We start out with a cause: exercising more. From that cause, we have three known effects that occurred within the group: weight loss, rise in confidence, and higher energy. Cause and effect are important terms to understand when conducting scientific experiments.

Physics

Nature of Motion

Cultures have been studying the movement of objects since ancient times. These studies have been prompted by curiosity and sometimes by necessity. On Earth, items move according to guidelines and have motion that is fairly predictable. To understand why an object moves along its path, it is important to understand what role forces have on influencing its movements. The term **force** describes an outside influence on an object. Force does not have to refer to something imparted by another object. Forces can act upon objects by touching them with a push or a pull, by friction, or without touch like a magnetic force or even gravity. Forces can affect the motion of an object.

To study an object's motion, it must be located and described. When locating an object's position, it can help to pinpoint its location relative to another known object. Comparing an object with respect to a known object is referred to as **establishing a frame of reference**. If the placement of one object is known, it is easier to locate another object with respect to the position of the original object.

Motion can be described by following specific guidelines called **kinematics**. Kinematics use mechanics to describe motion without regard to the forces that are causing such motions. Specific equations can be used when describing motions; these equations use time as a frame of reference. The equations are based on the change of an object's position (represented by x), over a change in time (represented by Δt). This describes an object's velocity, which is measured in meters/second (m/s) and described by the following equation:

$$v = \frac{\Delta x}{\Delta t} = \frac{x_f - x_i}{\Delta t}$$

Velocity is a **vector quantity**, meaning it measures the magnitude (how much) and the direction that the object is moving. Both of these components are essential to understanding and predicting the motion of objects. The scientist Isaac Newton did extensive studies on the motion of objects on Earth and came up with three primary laws to describe motion:

Law 1: An object in motion tends to stay in motion unless acted upon by an outside force. An object at rest tends to stay at rest unless acted upon by an outside force (also known as the **law of inertia**).

For example, if a book is placed on a table, it will stay there until it is moved by an outside force.

Law 2: The force acting upon an object is equal to the object's mass multiplied by its acceleration (also known as **F = ma**).

For example, the amount of force acting on a bug being swatted by a flyswatter can be calculated if the mass of the flyswatter and its acceleration are known. If the mass of the flyswatter is 0.3 kg and the acceleration of its swing is 2.0 m/s^2, the force of its swing can be calculated as follows:

$$m = 0.3 \, kg$$
$$a = 2.0 \, m/s^2$$
$$F = m \times a$$
$$F = (0.3) \times (2.0)$$
$$F = 0.6 \, N$$

Law 3: For every action, there is an equal and opposite reaction.

For example, when a person claps their hands together, the right hand feels the same force as the left hand, as the force is equal and opposite.

Another example is if a car and a truck run head-on into each other, the force experienced by the truck is equal and opposite to the force experienced by the car, regardless of their respective masses or velocities. The ability to withstand this amount of force is what varies between the vehicles and creates a difference in the amount of damage sustained.

Newton used these laws to describe motion and derive additional equations for motion that could predict the position, velocity, acceleration, or time for objects in motion in one and two dimensions. Since all of Newton's work was done on Earth, he primarily used Earth's gravity and the behavior of falling objects to design experiments and studies in free fall (an object subject to Earth's gravity while in flight). On Earth, the acceleration due to the force of gravity is measured at 9.8 meters per second^2 (m/s^2). This value is the same for anything on the Earth or within Earth's atmosphere.

Acceleration

Acceleration is the change in velocity over the change in time. It is given by the following equation:

$$a = \frac{\Delta v}{\Delta t} = \frac{v_f - v_i}{\Delta t}$$

Since velocity is the change in position (displacement) over a change in time, it is necessary for calculating an acceleration. Both of these are vector quantities, meaning they have magnitude and direction (or some amount in some direction). Acceleration is measured in units of distance over time^2 (meters/second^2 or m/s^2 in metric units).

For example, what is the acceleration of a vehicle that has an initial velocity of 35 m/s and a final velocity of 10 m/s over 5.0 s?

Using the givens and the equation:
$$a = \frac{\Delta v}{\Delta t} = \frac{v_f - v_i}{\Delta t}$$

V_f = 10 m/s

V_i = 35 m/s

Δt = 5.0 s

$$a = \frac{10 - 35}{5.0} = \frac{-25}{5.0} = -5.0 \, m/s^2$$

The vehicle is decelerating at –5.0 m/s^2.

If an object is moving with a constant velocity, its velocity does not change over time. Therefore, it has no (or 0) acceleration.

It is common to misuse vector terms in spoken language. For example, people frequently use the term "speed" in situations where the correct term would be "velocity." However, the difference between velocity and speed is not just that velocity must have a direction component with it. Average velocity and average speed actually are looking at two different distances as well. Average speed is calculated simply by dividing the total distance covered by the time it took to travel that distance. If someone runs four miles along a straight road north and then makes a 90-degree turn to the right and runs another three miles down that straight road east (such that the runner's route of seven miles makes up two sides of a rectangle) in seventy minutes, the runner's average speed was 6 miles per hour (one mile covered every ten minutes). Using the same course, the runner's average velocity would be about 4.29 miles per hour northeast.

Why is the magnitude less in the case of velocity? Velocity measures the change in position, or *displacement,* which is the shortest line between the starting point and ending point. Even if the path between these two points is serpentine or meanders all over the place racking up a great distance, the displacement is still just the shortest straight path between the change in the position of the object. In the case of the runner, the "distance" used to calculate velocity (the displacement) is the hypotenuse of the triangle that would connect the two side lengths at right angles to one another. Using basic trigonometric ratios, we know this distance is 5 miles (since the lengths of the other two legs are 3 miles and 4 miles and the Pythagorean Theorem says that $a^2 + b^2 = c^2$). Therefore, although the distance the runner covered was seven miles, his or her displacement was only five miles. Average velocity is thus calculated by taking the total time (70 minutes) and dividing it by the displacement (5 miles northeast). Therefore, to calculate average velocity of the runner, 70 minutes is divided by 5 miles, so each mile of displacement to the northeast was covered in 14 minutes. To find this rate in miles per hour, 60 minutes is divided by 14, to get 4.29 miles per hour northeast.

Another misconception is if something has a negative acceleration, it must be slowing down. If the change in position of the moving object is in a negative direction, it could have a negative velocity. If the acceleration is in the same direction as this negative velocity, it would be increasing the velocity in the negative direction, thus resulting in the object actually increasing in velocity.

For example, if west is designated to be a negative direction, a car increasing in speed to the west would have a negative velocity. Since it is increasing in speed, it would be accelerating in the negative direction, resulting in a negative acceleration.

Another common misconception is if a person is running around an oval track at a constant velocity, they would have no (or 0) acceleration because there is no change in the runner's velocity. This idea is incorrect because the person is changing direction the entire time they are running around the track, so there would be a change in their velocity, therefore; the runner would have an acceleration.

One final point regarding acceleration is that it can result from the force a rotating body exerts toward its center. For planets and other massive bodies, it is called **gravity**. This type of acceleration can also be utilized to separate substances, as in a centrifuge.

Projectile Motion

When objects are launched or thrown into the air, they exhibit what is called **projectile motion**. This motion takes a parabolic (or arced) path as the object rises and/or falls with the effect of gravity. In sports, if a ball is thrown across a field, it will follow a path of projectile motion. Whatever angle the object leaves the horizon is the same angle with which it will return to the horizon. The height the object achieves is referred to as the **y-component**, and the distance along the horizon the object travels is referred to as the **x-component.** To maximize the horizontal distance an object travels, it should be launched at a 45-degree angle relative to the horizon.

The following shows the range, or x-distance, an object can travel when launched at various angles:

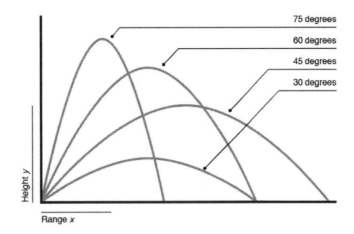

If something is traveling through the air without an internal source of power or any extra external forces acting upon it, it will follow these paths. All projectiles experience the effects of gravity while on Earth; therefore, they will experience a constant acceleration of 9.8 m/s^2 in a downward direction. While on its path, a projectile will have both a horizontal and vertical component to its motion and velocity. At the launch, the object has both vertical and horizontal velocity. As the object increases in height, the y-component of its velocity will diminish until the very peak of the object's path. At the peak, the y-component of the velocity is zero. Since the object still has a horizontal, or x-component, to its velocity, it would continue its motion. There is also a constant acceleration of 9.8 m/s^2 acting in the downward direction on the object the entire time, so at the peak of the object's height, the velocity would then switch from zero to being pulled down with the acceleration. The entire path of the object takes a specific amount of time, based on the initial launch angle and velocity. The time and distance traveled can be calculated using the kinematic equations of motion. The time it takes the object to reach its maximum height is exactly half of the time for the object's entire flight.

Similar motion is exhibited if an object is thrown from atop a building, bridge, cliff, etc. Since it would be starting at its maximum height, the motion of the object would be comparable to the second half of the path in the above diagram.

Friction

As previously stated with Newton's laws, forces act upon objects on the Earth. If an object is resting on a surface, the effect of gravity acting upon its mass produces a force referred to as weight. This weight touches the surface it is resting on, and the surface produces a normal force perpendicular to this surface. If an outside force acts upon this object, its movement will be resisted by the surfaces rubbing on each other.

Friction is the term used to describe the force that opposes motion, or the force experienced when two surfaces interact with each other. Every surface has a specific amount with which it resists motion, called a **coefficient of friction.** The coefficient of friction is a proportion calculated from the force of friction divided by the **normal force** (force produced perpendicular to a surface).

$$\mu_s = \frac{F_s}{F_N}$$

There are different types of friction between surfaces. If something is at rest, it has a **static** (non-moving) friction. It requires an outside force to begin its movement. The coefficient of static friction for that material multiplied by the normal force would need to be greater than the force of static friction to get the object moving. Therefore, the force required to move an object must be greater than the force of static friction:

$$F_s \leq \mu_s \times F_N$$

Once the object is in motion, the force required to maintain this movement only needs to be equivalent to the value of the force of kinetic (moving) friction. To calculate the force of **kinetic** friction, simply multiply the coefficient of kinetic friction for that surface by the normal force:

$$F_k = \mu_k \times F_N$$

The force required to start an object in motion is larger than the force required to continue its motion once it has begun:

$$F_s \geq F_k$$

Friction not only occurs between solid surfaces; it also occurs in air and liquids. In air, it is called **air resistance,** or drag, and in water, it is called **viscosity**.

For example, what would the coefficient of static friction be if a 5.0 N force was applied to push a 20 kg crate, from rest, across a flat floor?

First, the normal force could be found to counter the force from the weight of the object, which would be the mass multiplied by gravity:

$$F_N = mass \times gravity$$

$$F_N = 20 \, kg \times 9.8 \, \frac{m}{s^2}$$

$$F_N = 196 \, N$$

Next, the coefficient of static friction could be found by dividing the frictional force by the normal force:

$$\mu_s = \frac{F_s}{F_N}$$

$$\mu_s = \frac{5.0 \, N}{196 \, N}$$

$$\mu_s = 0.03$$

Since it is a coefficient, the units cancel out, so the solution is unitless. The coefficient of static friction should also be less than 1.0.

Rotation

An object moving on a circular path has **momentum** (a measurement of an object's mass and velocity in a direction); for circular motion, it is called **angular momentum,** and this value is determined by rotational inertia, rotational velocity, and the distance of the mass from the axis of rotation, or the center of rotation.

If objects are exhibiting circular motion, they are demonstrating the conservation of angular momentum. The angular momentum of a system is always constant, regardless of the placement of the mass. As stated above, the rotational inertia of an object can be affected by how far the mass of the object is placed with respect to the center of rotation (axis of rotation). A larger distance between the mass and the center of rotation means a slower rotational velocity. Reversely, if a mass is closer to the center of rotation, the rotational velocity increases. A change in the placement of the mass affects the value of the rotational velocity, thus conserving the angular momentum. This is true as long as no external forces act upon the system.

For example, if an ice skater is spinning on one ice skate and extends his or her arms out (or increases the distance between the mass and the center of rotation), a slower rotational velocity is created. When the skater brings the arms in close to the body (or lessens the distance between the mass and the center of rotation), his or her rotational velocity increases, and he or she spins much faster. Some skaters extend their arms straight up above their head, which causes the axis of rotation to extend, thus removing any distance between the mass and the center of rotation and maximizing the rotational velocity.

Consider another example. If a person is selecting a horse on a merry-go-round, the placement of their selection can affect their ride experience. All the horses are traveling with the same rotational speed, but to travel along the same plane as the merry-go-round, a horse closer to the outside will have a greater linear speed due to being farther away from the axis of rotation. Another way to think of this is that an outside horse must cover more distance than a horse near the inside to keep up with the rotational speed of the merry-go-round platform. Based on this information, thrill seekers should always select an outer horse to experience a greater linear speed.

Uniform Circular Motion

When an object exhibits circular motion, its motion is centered around an axis. An **axis** is an invisible line on which an object can rotate. This type of motion is most easily observed on a toy top. There is actually a point (or rod) through the center of the top on which the top can be observed to be spinning. This is also called the axis. An axis is the location about which the mass of an object or system would rotate if free to spin.

In the instance of utilizing a lever to lift an object, it can be helpful to calculate the amount of force needed at a specific distance, applied perpendicular to the axis of motion, to calculate the torque, or circular force, necessary to move something. This is also employed when using a wrench to loosen a bolt. The equation for calculating the force in a circular direction, or perpendicular to an axis, is as follows:

$$Torque = F_{\perp} \times distance\ of\ lever\ arm\ from\ the\ axis\ of\ rotation$$

$$\tau = F_{\perp} \times d$$

For example, what torque would result from a 20 N force being applied to a lever 5 meters from its axis of rotation?

$$\tau = 20\,N \times 5\,m$$

$$\tau = 100\,N \times m$$

The amount of torque would be 100 N×m. The units would be Newton meters because it is a force applied at a distance away from the axis of rotation.

When objects move in a circle by spinning on their own axis, or because they are tethered around a central point (also considered an axis), they exhibit circular motion. Circular motion is similar in many ways to linear (straight line) motion; however, there are some additional facts to note. When an object spins or rotates on or around an axis, a force that feels like it is pushing out from the center of the circle is created. The force is actually pulling into the center of the circle. A reactionary force is what is creating the feeling of pushing out. The inward force is the real force, and this is called **centripetal force**. The outward, or reactionary, force is called **centrifugal force**. The reactionary force is not the real force; it just feels like it is there. This can also be referred to as a **fictional force.** The true force is the one pulling inward, or the centripetal force. The terms centripetal and centrifugal are often mistakenly interchanged.

For example, the method a traditional-style washing machine uses to spin a load of clothes to expunge the water from the load is to spin the machine barrel in a circle at a high rate of speed. During this spinning, the centripetal force is pulling in toward the center of the circle. At the same time, the reactionary force to the centripetal force is pressing the clothes up against the outer sides of the barrel, which expels the water out of the small holes that line the outer wall of the barrel.

Kinetic Energy and Potential Energy

There are two main types of energy. The first type is called **potential energy** (or gravitational potential energy), and it is stored energy, or energy due to an object's height from the ground.

The second type is called **kinetic energy**. Kinetic energy is the energy of motion. If an object is moving, it will have some amount of kinetic energy.

For example, if a roller-coaster car is sitting on the track at the top of a hill, it would have all potential energy and no kinetic energy. As the roller coaster travels down the hill, the energy converts from potential energy into kinetic energy. At the bottom of the hill, where the car is traveling the fastest, it would have all kinetic energy and no potential energy.

Another measure of energy is the **total mechanical energy** in a system. This is the sum (or total) of the potential energy plus the kinetic energy of the system. The total mechanical energy in a system is always conserved. The amounts of the potential energy and kinetic energy in a system can vary, but the total mechanical energy in a situation would remain the same.

The equation for the mechanical energy in a system is as follows:

$$ME = PE + KE$$

$$(Mechanical\ Energy\ =\ Potential\ Energy\ +\ Kinetic\ Energy)$$

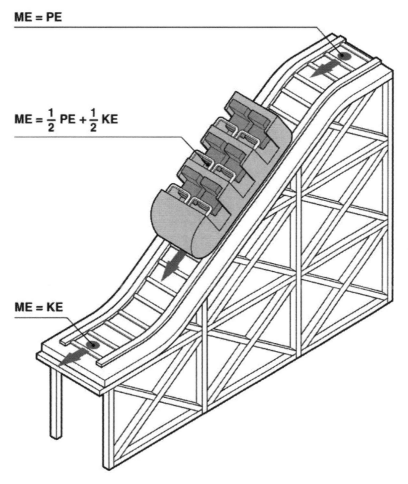

Energy can transfer or change forms, but it cannot be created or destroyed. This transfer can take place through waves (including light waves and sound waves), heat, impact, etc.

There is a fundamental law of thermodynamics (the study of heat and movement) called **conservation of energy.** This law states that energy cannot be created or destroyed, but rather energy is transferred to different forms involved in a process. For instance, a car pushed beginning at one end of a street will not continue down that street forever; it will gradually come to a stop some distance away from where it was originally pushed. This does not mean the energy has disappeared or has been exhausted; it means the energy has been transferred to different mediums surrounding the car. Some of the energy is dissipated by the frictional force from the road on the tires, the air resistance from the movement of the car, the sound from the tires on the road, and the force of gravity pulling on the car. Each value can be calculated in a number of ways, including measuring the sound waves from the tires, the temperature change in the tires, the distance moved by the car from start to finish, etc. It is important to understand that many processes factor into such a small situation, but all situations follow the conservation of energy.

215

Just like the earlier example, the roller coaster at the top of a hill has a measurable amount of potential energy; when it rolls down the hill, it converts most of that energy into kinetic energy. There are still additional factors such as friction and air resistance working on the coaster and dissipating some of the energy, but energy transfers in every situation.

There are six basic machines that utilize the transfer of energy to the advantage of the user. These machines function based on an amount of energy input from the user and accomplish a task by distributing the energy for a common purpose. These machines are called **simple machines** and include the lever, pulley, wedge, inclined plane, screw, and wheel and axle.

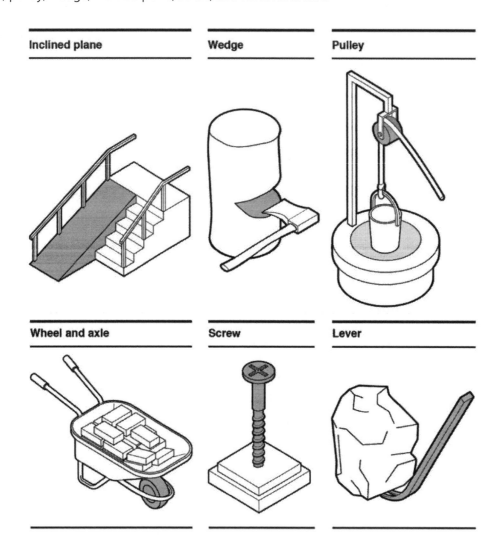

The use of simple machines can help by requiring less force to perform a task with the same result. This is referred to as a **mechanical advantage**.

For example, if a father is trying to lift his child into the air with his arms to pick an apple from a tree, it would require less force to place the child on one end of a teeter totter and push the other end of the teeter totter down to elevate the child to the same height to pick the apple. In this example, the teeter totter is a lever.

Linear Momentum and Impulse

The motion of an object can be expressed as momentum. This is a calculation of an object's mass times its velocity. **Momentum** can be described as the amount an object will continue moving along its current course. Momentum in a straight line is called **linear momentum**. Just as energy can be transferred and conserved, so can momentum.

Momentum is denoted by the letter p and calculated by multiplying an object's mass by its velocity.

$$p = m \times v$$

For example, if a car and a truck are moving at the same velocity (25 meters per second) down a highway, they will not have the same momentum because they do not have the same mass. The mass of the truck (3500 kg) is greater than that of the car (1000 kg); therefore, the truck will have more momentum. In a head-on collision, the truck's momentum is greater than the car's, and the truck will cause more damage to the car than the car will to the truck. The equations to compare the momentum of the car and the truck are as follows:

$$p_{truck} = mass_{truck} \times velocity_{truck} \qquad\qquad p_{car} = mass_{car} \times velocity_{car}$$

$$p_{truck} = 3500 \; kg \times 25 \; m/s \qquad\qquad p_{car} = 1000 \; kg \times 25 \; m/s$$

$$p_{truck} = 87{,}500 \; N \qquad\qquad p_{car} = 25{,}000 \; N$$

The momentum of the truck is greater than that of the car.

The amount of force during a length of time creates an impulse. This means if a force acts on an object during a given amount of time, it will have a determined impulse. However, if the length of time can be extended, the force will be less due to the conservation of momentum.

For a car crash, the total momentum of each car before the collision would need to equal the total momentum of the cars after the collision. There are two main types of collisions: elastic and inelastic. For the example with a car crash, in an elastic collision, the cars would be separate before the collision, and they would remain separated after the collision. In the case of an inelastic collision, the cars would be separate before the collision, but they would be stuck together after the collision. The only difference would be in the way the momentum is calculated.

For elastic collisions:

$$total \; momentum_{before} = total \; momentum_{after}$$

$$(mass_{car\,1} \times velocity_{car\,1}) + (mass_{car\,2} \times velocity_{car\,2}) = (mass_{car\,1} + mass_{car\,2}) \times velocity_{car\,1 \; \& \; car\,2}$$

The damage from an impact can be lessened by extending the time of the actual impact. This is called the measure of the impulse of a collision. It can be calculated by multiplying the change in momentum by the amount of time involved in the impact.

$$I = change \; in \; momentum \times time$$

$$I = \Delta p \times time$$

If the time is extended, the force (or change in momentum) is decreased. Conversely, if the time is shortened, the force (or change in momentum) is increased. For example, when catching a fast baseball, it helps soften the blow of the ball to follow through, or cradle the catch. This technique is simply extending the time of the application of the force of the ball, so the impact of the ball does not hurt the hand.

For example, if martial arts experts want to break a board by executing a chop from their hands, they need to exert a force on a small point on the board, extremely quickly. If they slow down the time of the impact from the force of their hands, they will probably injure their striking hand and not break the board.

Often, law enforcement officials will use rubber bullets instead of regular bullets to apprehend a criminal. The benefit of the rubber bullet is that the elastic material of the bullet bounces off the target but hits the target with nearly the same momentum as a regular bullet. Since the length of time the rubber bullet is in contact with the target is decreased, the amount of force from the bullet is increased. This method can knock a subject off their feet by the large force and the short time of the impact without causing any lasting harm to the individual. The difference in the types of collisions is noted through the rubber bullet bouncing off the individual, so both the bullet and the subject are separate before the collision and separate after the collision. With a regular bullet, the bullet and subject are separate before the collision, but a regular bullet would most likely not be separated by the subject after the collision.

Universal Gravitation

Every object in the universe that has mass causes an attractive force to every other object in the universe. The amount of attractive force depends on the masses of the two objects in question and the distance that separates the objects. This is called the **law of universal gravitation** and is represented by the following equation:

$$F = G \frac{m_1 m_2}{r^2}$$

In this equation, the force, F, between two objects, m_1 and m_2, is indirectly proportional to the square of the distance separating the two objects. A general gravitational constant G ($6.67 \times 10^{-11} \frac{N \cdot m^2}{kg^2}$) is multiplied by the equation. This constant is quite small, so for the force between two objects to be noticeable, they must have sizable masses.

To better understand this on a large scale, a prime representation could be viewed by satellites (planets) in the solar system and the effect they have on each other. All bodies in the universe have an attractive force between them. This is closely seen by the relationship between the Earth and the moon. The Earth and the moon both have a gravitational attraction that affects each other. The moon is smaller in mass than the Earth; therefore, it will not have as big of an influence as the Earth has on it. The attractive force from the moon is observed by the systematic push and pull on the water on the face of the Earth by the rotations the moon makes around the Earth. The tides in oceans and lakes are caused by the moon's gravitational effect on the Earth. Since the moon and the Earth have an attractive force between them, the moon pulls on the side of the Earth closest to the moon, causing the waters to swell (high tide) on that side and leave the ends 90 degrees away from the moon, causing a low tide there. The water on the side of the Earth farthest from the moon experiences the least amount of gravitational attraction so it collects on that side in a high tide.

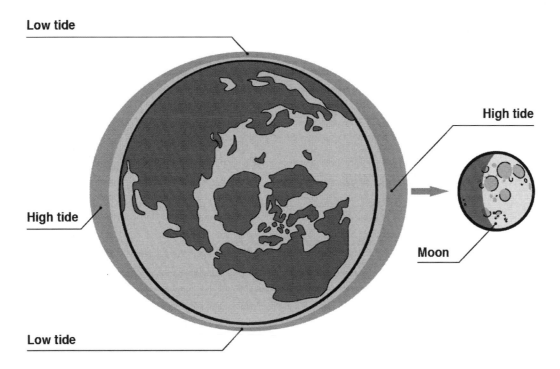

The universal law of gravitation is taken primarily from the works of Johannes Kepler and his laws of planetary motion. These include the fact that the paths of the orbits of the planets are not perfect circles, but ellipses, around the sun. The area swept out between the planet and the sun is equal at every point in the orbit due to fluctuation in speed at different distances. Finally, the period (*T*) of a planet's motion squared is inversely proportional to the distance (*r*) between that planet and the sun cubed.

$$\frac{T^2}{r^3}$$

Sir Isaac Newton used this third law and applied it to the idea of forces and their effects on objects. The effect of the gravitational forces of the moon on the Earth are noted in the tides, and the effect of the forces of the Earth on the moon are noted in the fact that the moon is caught in an orbit around the earth. Since the moon is traveling at a velocity tangent to its orbit around the Earth and the Earth keeps attracting it in, the moon does not escape and does not crash into the Earth. The moon will continue this course due to the attractive gravitational force between the Earth and the moon. Albert Einstein later applied Newton's adaptation of Kepler's laws. Einstein was able to develop a more advanced theory, which could explain the motions of all the planets and even be applied beyond the solar system. These theories have also been beneficial for predicting behaviors of other objects in the Earth's atmosphere, such as shuttles and astronauts.

Waves and Sound

Mechanical waves are a type of wave that pass through a medium (solid, liquid, or gas). There are two basic types of mechanical waves: longitudinal and transverse.

A **longitudinal wave** has motion that is parallel to the direction of the wave's travel. It can best be shown by compressing one side of a tethered spring and then releasing that end. The movement travels in a bunching and then unbunching motion, across the length of the spring and back until the energy is dissipated through noise and heat.

A transverse wave has motion that is perpendicular to the direction of the wave's travel. The particles on a transverse wave do not move across the length of the wave but oscillate up and down to create the peaks and troughs observed on this type of wave.

A wave with a mix of both longitudinal and transverse motion can be seen through the motion of a wave on the ocean, with peaks and troughs, oscillating particles up and down.

Mechanical waves can carry energy, sound, and light. Mechanical waves need a medium through which transport can take place. However, an electromagnetic wave can transmit energy without a medium, or in a vacuum.

Sound travels in waves and is the movement of vibrations through a medium. It can travel through air (gas), land, water, etc. For example, the noise a human hears in the air is the vibration of the waves as they reach the ear. The human brain translates the different frequencies (pitches) and intensities of the vibrations to determine what created the noise.

A tuning fork has a predetermined frequency because of the size (length and thickness) of its tines. When struck, it allows vibrations between the two tines to move the air at a specific rate. This creates a specific tone (or note) for that size of tuning fork. The number of vibrations over time is also steady for that tuning fork and can be matched with a frequency (the number of occurrences over time). All sounds heard by the human ear are categorized by using frequency and measured in **hertz** (the number of cycles per second).

The intensity (or loudness) of sound is measured on the Bel scale. This scale is a ratio of one sound's intensity with respect to a standard value. It is a logarithmic scale, meaning it is measured by factors of ten. But the value that is 1/10 of this value, the decibel, is the measurement used more commonly for the intensity of pitches heard by the human ear.

The Doppler effect applies to situations with both light and sound waves. The premise of the Doppler effect is that, based on the relative position or movement of a source and an observer, waves can seem shorter or longer than they are. When the Doppler effect is experienced with sound, it warps the noise being heard by the observer by making the pitch or frequency seem shorter or higher as the source is approaching and then longer or lower as the source is getting farther away. The frequency and pitch of the source never actually change, but the sound in respect to the observer's position makes it seem like the sound has changed. This effect can be observed when an emergency siren passes by an observer on the road. The siren sounds much higher in pitch as it approaches the observer and then lower after it passes and is getting farther away.

The Doppler effect also applies to situations involving light waves. An observer in space would see light approaching as being shorter wavelengths than it was, causing it to appear blue. When the light wave is getting farther away, the light would look red due to the apparent elongation of the wavelength. This is called the **red-blue shift.**

A recent addition to the study of waves is the gravitational wave. Its existence has been proven and verified, yet the details surrounding its capabilities are still under inquiry. Further understanding of gravitational waves could help scientists understand the beginnings of the universe and how the existence of the solar system is possible. This understanding could also include the future exploration of the universe.

Light

The movement of light is described like the movement of waves. Light travels with a wave front and has an amplitude (a height measured from the neutral), a cycle or wavelength, a period, and energy. Light travels at approximately 3.00×10^8 m/s and is faster than anything created by humans.

Light is commonly referred to by its measured **wavelengths**, or the length for it to complete one cycle. Types of light with the longest wavelengths include radio, TV, micro, and infrared waves. The next set of wavelengths are detectable by the human eye and make up the visible spectrum. The visible spectrum has wavelengths of 10^{-7} m, and the colors seen are red, orange, yellow, green, blue, indigo, and violet. Beyond the visible spectrum are even shorter wavelengths (also called the **electromagnetic spectrum**) containing ultraviolet light, x-rays, and gamma rays. The wavelengths outside of the visible light range can be harmful to humans if they are directly exposed, especially for long periods of time.

For example, the light from the sun has a small percentage of ultraviolet (UV) light, which is mostly absorbed by the UV layer of the Earth's atmosphere. When this layer does not filter out the UV rays, the exposure to the wavelengths can be harmful to human skin. When there is an extra layer of pollutant and the light from the sun is trapped by repeated reflection back to the ground, unable to bounce back into space, it creates another harmful condition for the planet called the **greenhouse effect**. This is an overexposure to the sun's light and contributes to global warming by increasing the temperatures on Earth.

When a wave crosses a boundary or travels from one medium to another, certain actions take place. If the wave travels through one medium into another, it experiences **refraction**, which is the bending of the wave from one medium's density to another, altering the speed of the wave.

For example, a side view of a pencil in half a glass of water appears as though it is bent at the water level. What the viewer is seeing is the refraction of light waves traveling from the air into the water. Since the wave speed is slowed in water, the change makes the pencil appear bent.

When a wave hits a medium that it cannot pass through, it is bounced back in an action called **reflection**. For example, when light waves hit a mirror, they are reflected, or bounced off, the back of the mirror. This can cause it to seem like there is more light in the room due to the doubling back of the initial wave. This is also how people can see their reflection in a mirror.

When a wave travels through a slit or around an obstacle, it is known as **diffraction**. A light wave will bend around an obstacle or through a slit and cause a diffraction pattern. When the waves bend around an obstacle, it causes the addition of waves and the spreading of light on the other side of the opening.

Optics

The dispersion of light describes the splitting of a single wave by refracting its components into separate parts. For example, if a wave of white light is sent through a dispersion prism, the light wave appears as its separate rainbow-colored components due to each colored wavelength being refracted in the prism.

Different things occur when wavelengths of light hit boundaries. Objects can absorb certain wavelengths of light and reflect others, depending on the boundaries. This becomes important when an object appears to be a certain color. The color of the object is not actually within the makeup of that object, but by what wavelengths are being transmitted by that object. For example, if a table appears to be red, that means the table is absorbing all wavelengths of visible light except those of the red wavelength. The table is reflecting, or transmitting, the wavelengths associated with red back to the human eye, and therefore, the table appears red.

Interference describes when an object affects the path of a wave or another wave interacts with that wave. Waves interacting with each other can result in either constructive interference or destructive interference based on their positions. For constructive interference, the waves are in sync and combine to reinforce each other. In the case of deconstructive interference, the waves are out of sync and reduce the effect of each other to some degree. In scattering, the boundary can change the direction or energy of a wave, thus altering the entire wave. Polarization changes the oscillations of a wave and can alter its appearance in light waves. For example, polarized sunglasses take away the "glare" from sunlight by altering the oscillation pattern observed by the wearer.

When a wave hits a boundary and is completely reflected back or cannot escape from one medium to another, it is called **total internal reflection**. This effect can be seen in a diamond with a brilliant cut. The angle cut on the sides of the diamond causes the light hitting the diamond to be completely reflected back inside the gem and makes it appear brighter and more colorful than a diamond with different angles cut into its surface.

When reflecting light, a mirror can be used to observe a virtual (not real) image. A plane mirror is a piece of glass with a coating in the background to create a reflective surface. An image is what the human eye sees when light is reflected off the mirror in an unmagnified manner. If a curved mirror is used for reflection, the image seen will not be a true reflection, but will either be magnified or made to appear smaller than its actual size. Curved mirrors can also make an object appear closer or farther away than its actual distance from the mirror.

Lenses can be used to refract or bend light to form images. Examples of lenses are human eye, microscopes, and telescopes. The human eye interprets the refraction of light into images that humans understand to be actual size. When objects are too small to be observed by the unaided human eye, microscopes allow the objects to be enlarged enough to be seen. Telescopes allow objects that are too far away to be seen by the unaided eye to be viewed. Prisms are pieces of glass that can have a wavelength of light enter one side and appear to be broken down into its component wavelengths on the other side, due to the slowing of certain wavelengths within the prism, more than other wavelengths.

The Nature of Electricity

Electrostatics is the study of electric charges at rest. A balanced atom has a neutral charge from its number of electrons and protons. If the charge from its electrons is greater than or less than the charge of its protons, the atom has a charge. If the atom has a greater charge from the number of electrons than protons, it has a negative charge. If the atom has a lesser charge from the number of electrons than protons, it has a positive charge. Opposite charges attract each other, while like charges repel each other, so a negative attracts a positive, and a negative repels a negative. Similarly, a positive charge repels a positive charge. Just as energy cannot be created or destroyed, neither can charge; charge can only be transferred. The transfer of charge can occur through touch, or the transfer of electrons. Once electrons have transferred from one object to another, the charge has been transferred.

For example, if a person wears socks and scuffs their feet across a carpeted floor, the person is transferring electrons to the carpeting through the friction from his or her feet. Additionally, if that person then touches a light switch, they he or she receive a small shock. This "shock" is the person feeling the electrons transferring from the switch to their hand. Since the person lost electrons to the carpet, that person now has fewer negative charges, resulting in a net positive charge. Therefore, the electrons from the light switch are attracted to the person for the transfer. The shock felt is the electrons moving from the switch to the person's finger.

Another method of charging an object is through induction. **Induction** occurs when a charged object is brought near two touching stationary objects. The electrons in the objects will attract and cluster near another positively-charged object and repel away from a negatively-charged object held nearby. The stationary objects will redistribute their electrons to allow the charges to reposition themselves closer or farther away. This redistribution will cause one of the touching stationary objects to be negatively charged and the other to be positively charged. The overall charges contained in the stationary objects remain the same but are repositioned between the two objects.

Another way to charge an object is through **polarization**. Polarization can occur simply by the reconfiguration of the electrons within a single object.

For example, if a girl at a birthday party rubs a balloon on her hair, the balloon could then cling to a wall if it were brought close enough. This would be because rubbing the balloon causes it to become negatively charged. When the balloon is held against a neutrally-charged wall, the negatively charged balloon repels all the wall's electrons, causing a positively-charged surface on the wall. This type of charge is temporary, due to the massive size of the wall, and the charges will quickly redistribute.

An electric current is produced when electrons carry charge across a length. To make electrons move so they can carry this charge, a change in voltage must be present. On a small scale, this is demonstrated through the electrons traveling from the light switch to a person's finger in the example where the person had run their socks on a carpet. The difference between the charge in the switch and the charge in the finger causes the electrons to move. On a larger and more sustained scale, this movement would need to be more controlled. This can be achieved through batteries/cells and generators. Batteries or cells have a chemical reaction that takes place inside, causing energy to be released and charges to move freely. Generators convert mechanical energy into electric energy for use after the reaction.

For example, if a wire runs touching the end of a battery to the end of a lightbulb, and then another wire runs touching the base of the lightbulb to the opposite end of the original battery, the lightbulb will light up. This is due to a complete circuit being formed with the battery and the electrons being carried across the voltage drop (the two ends of the battery). The appearance of the light from the bulb is the visible presence of the electrons in the filament of the bulb.

Electric energy can be derived from a number of sources, including coal, wind, sun, and nuclear reactions. Electricity has numerous applications, including being transferable into light, sound, heat, or magnetic forces.

Magnetism and Electricity

Magnetic forces occur naturally in specific types of materials and can be imparted to other types of materials. If two straight iron rods are observed, they will naturally have a negative end (pole) and a positive end (pole). These charged poles follow the rules of any charged item: Opposite charges attract, and like charges repel. When set up positive to negative, they will attract each other, but if one rod is turned around, the two rods will repel each other due to the alignment of negative to negative poles and positive to positive poles. When poles are identified, magnetic fields are observed between them.

If small iron filings (a material with natural magnetic properties) are sprinkled over a sheet of paper resting on top of a bar magnet, the field lines from the poles can be seen in the alignment of the iron filings, as pictured below:

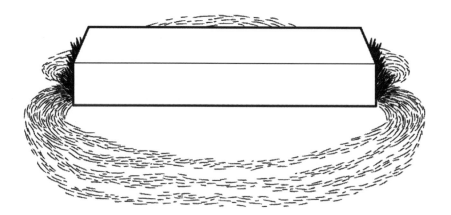

These fields naturally occur in materials with magnetic properties. There is a distinct pole at each end of such a material. If materials are not shaped with definitive ends, the fields will still be observed through the alignment of poles in the material. For example, a circular magnet does not have ends but still has a magnetic field associated with its shape, as pictured below:

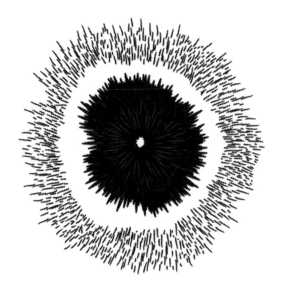

Magnetic forces can also be generated and amplified by using an electric current. For example, if an electric current is sent through a length of wire, it creates an electromagnetic field around the wire from the charge of the current. This force is from the moving of negatively-charged electrons from one end of the wire to the other. This is maintained as long as the flow of electricity is sustained. The magnetic field can also be used to attract and repel other items with magnetic properties. A smaller or larger magnetic force can be generated around this wire, depending on the strength of the current in the wire. As soon as the current is stopped, the magnetic force also stops.

Magnetic energy can be harnessed, or manipulated, from natural sources or from a generated source (a wire carrying electric current). When a core with magnetic properties (such as iron) has a wire wrapped around it in circular coils, it can be used to create a strong, non-permanent electromagnet. If current is run through the wrapped wire, it generates a magnetic field by polarizing the ends of the metal core, as described above, by moving the negative charge from one end to the other. If the direction of the current is reversed, so is the direction of the magnetic field due to the poles of the core being reversed. The term **non-permanent** refers to the fact that the magnetic field is generated only when the current is present, but not when the current is stopped. The following is a picture of a small electromagnet made from an iron nail, a wire, and a battery:

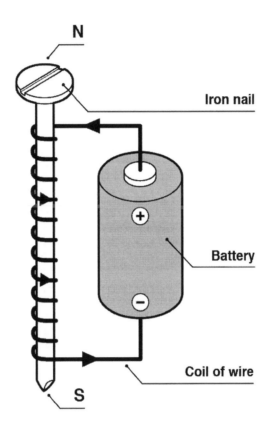

This type of electromagnetic field can be generated on a larger scale using more sizable components. This type of device is useful in the way it can be controlled. Rather than having to attempt to block a permanent magnetic field, the current to the system can simply be stopped, thus stopping the magnetic field. This provides the basis for many computer-related instruments and magnetic resonance imaging (MRI) technology. Magnetic forces are used in many modern applications, including the creation of super-speed transportation. Super magnets are used in rail systems and supply a cleaner form of energy than coal or gasoline. Another example of the use of super-magnets is seen in medical equipment, specifically MRI. These machines are highly sophisticated and useful in imaging the internal workings of the human body. For super-magnets to be useful, they often must be cooled down to extremely low temperatures to dissipate the amount of heat generated from their extended usage. This can be done by flooding the magnet with a super-cooled gas such as helium or liquid nitrogen. Much research is continuously done in this field to find new ceramic–metallic hybrid materials that have structures that can maintain their charge and temperature within specific guidelines for extended use.

Practice Questions

1. Which of the following is an example of a physical property of substances?
    a. Heat of combustion
    b. Reactivity
    c. Odor
    d. Toxicity
    e. Flammability

2. Which type of matter has molecules that cannot move within the substance and breaks evenly across a plane due to the symmetry of its molecular arrangement?
    a. Gases
    b. Elements
    c. Liquids
    d. Amorphous solids
    e. Crystalline solids

3. What type of reactions involve the breaking and re-forming of bonds between atoms?
    a. Chemical
    b. Physical
    c. Isotonic
    d. Electron
    e. Nuclear

4. If a vehicle increases speed from 20 m/s to 40 m/s over a time period of 20 seconds, what is the vehicle's rate of acceleration?
    a. 2 m/s
    b. 1 m/s^2
    c. 10 m/s
    d. 10 m/s^2
    e. 1 m/s

5. Which concept explains the rising of tides in the ocean?
    a. The sun pulling the ocean water up.
    b. The Earth pushing its oceans towards the moon.
    c. The Earth's gravitational effect on the moon.
    d. The sun pushing the ocean waters up.
    e. The moon's gravitational effect pulling on the Earth.

6. What is friction?
    a. It is the normal force of an object that is perpendicular to the surface.
    b. It is the force required to maintain the movement of an object.
    c. It is the force that opposes motion.
    d. It is the velocity of an object.
    e. It is the force required to stop a moving object.

7. In which part of the atom is 99 percent of its mass found?
    a. Orbitals
    b. Nucleus
    c. Protons
    d. Neutrons
    e. Electrons

8. What is different between the isotopes of an atom?
    a. The number of protons
    b. The number of orbitals
    c. The number of neutrons
    d. The location of the nucleus
    e. The number of electrons

9. Where should a person sit on a merry-go-round to experience the greatest linear speed of the ride?
    a. Halfway between the axis and the outer most horse
    b. $\frac{1}{3}$ of the way between the axis and the outer most horse
    c. On the outermost horse
    d. On the innermost horse
    e. On the highest horse

10. Which type of reaction is represented by the following equation: A + B → C?
    a. Decomposition
    b. Combustion
    c. Enzymatic
    d. Combination
    e. Oxidation/Reduction

11. How do catalysts work to increase the rate of a reaction?
    a. By decreasing the number of reactants
    b. By lowering the activation energy required to drive the reaction forward
    c. By increasing the temperature of the reaction environment
    d. By decreasing the number of products that are formed
    e. By increasing the concentration of reactants so more collisions occur

12. What torque would result from a 35 N force that was applied to a lever 10 meters away from the axis of rotation?
    a. 35 N×m
    b. 0.35 N×m
    c. 10 N×m
    d. 3.5 N×m
    e. 350 N×m

13. What is the molar mass of a substance?
    a. The mass of one mole of the substance
    b. The mass of ten molecules of the substance
    c. The number of protons in the substance
    d. The number of atoms in the substance
    e. The mass of one molecule of the substance

14. Which of the following is one main type of transfer that occurs in redox reactions?
    a. Transfer of carbon
    b. Transfer of hydrogen
    c. Transfer of nitrogen
    d. Transfer of sulfur
    e. Transfer of helium

15. Which principle of light occurs when a wave travels through a slit or around an obstacle?
    a. Refraction
    b. Diffraction
    c. Reflection
    d. Greenhouse effect
    e. Interference

16. In the Periodic Table, what similarity do the elements in columns have with each other?
    a. They have the same atomic number
    b. They have similar chemical properties
    c. They have similar electron valence configurations
    d. They start with the same first letter
    e. They have the same density

17. Which is NOT an example of how to charge an object?
    a. Transferring electrons
    b. Induction
    c. Polarization
    d. Refraction
    e. Friction

18. What number on the pH scale indicates a neutral solution?
    a. 13
    b. 8
    c. 7
    d. 0
    e. 1

19. For what entity does kinematics provide a guideline?
    a. Motion
    b. Time
    c. Mass
    d. Position
    e. Force

20. Which answer choice describes Newton's law of inertia?
    a. For every action, there is an equal and opposite reaction.
    b. Velocity is equal to a change in position divided by the change in time.
    c. The force acting upon an object is equal to the object's mass multiplied by its acceleration.
    d. An object at rest tends to stay at rest and an object in motion tends to stay in motion unless an outside force acts upon it.
    e. Momentum, the product of an object's mass and its velocity, describes the degree to which the object will continue moving along its current path.

Answer Explanations

1. C: Physical properties of substances are those that can be observed without changing that substance's chemical composition, such as odor, color, density, and hardness. Heat of combustion, reactivity, flammability, and toxicity are all chemical properties of substances. They describe the way in which a substance may change into a different substance. They cannot be observed without changing the identity of the substance.

2. E: Solids have molecules that are packed together tightly and cannot move within their substance. Crystalline solids have atoms or molecules that are arranged symmetrically, making all of the bonds of even strength. When they are broken, they break along a plane of molecules, creating a straight edge. Amorphous solids do not have the symmetrical makeup of crystalline solids. They do not break evenly. Gases and liquids both have molecules that move around freely. Elements are not a distinct phase of matter; they are substances formed by only a single type of atom.

3. A: Chemical reactions are processes that involve the changing of one set of substances to a different set of substances. To accomplish this, the bonds between the atoms in the molecules of the original substances need to be broken. The atoms are rearranged, and new bonds are formed to make the new set of substances. Combination reactions involve two or more reactants becoming one product. Decomposition reactions involve one reactant becoming two or more products. Combustion reactions involve the use of oxygen as a reactant and generally include the burning of a substance. Choices *C* and *D* are not discussed as specific reaction types. Nuclear reactions, Choice *E*, involve a change in the nucleus of the atom rather than a rearrangement of electrons.

4. B: The equation for calculating acceleration is:

$$a = \frac{\Delta v}{\Delta t} = \frac{v_f - v_i}{\Delta t}$$

In this case, V_f = 40 m/s, V_i = 20 m/s, and Δt = 20 s. Plugging in the numbers gives an acceleration rate of 1 m/s per second, which is written as 1 m/s^2, Choice *B*.

5. E: The moon has a gravitational pull on the Earth as it orbits around it. The pull causes the tides in oceans to rise, and at the point where the moon is closest to the Earth is when high tide occurs. The Earth orbits around the sun but that does not affect the tides of the ocean; therefore, Choices *A* and *D* are incorrect. The tides are caused by the moon's pull on the ocean and not the other way around, so Choices *B* and *C* are incorrect.

6. C: Friction is the force of an object that opposes motion when it interacts with another surface. Every surface has some amount of friction that resists motion. All objects have a perpendicular force that they exert on surfaces they touch, Choice *A*, but this is not defined as friction; it is called normal force. Friction specifically involves the resisting force when trying to move an object and does not involve the velocity of an object; therefore, Choices *B*, *D*, and *E* are incorrect.

7. B: The nucleus contains the protons and neutrons of the atom. Together, they make up 99 percent of the atom's mass. Electrons, Choice *E*, are found in the orbitals, Choice *A*, surrounding the nucleus and do not have as much mass as the protons or neutrons. Protons and neutrons, Choices *C* and *D*, have similar masses, but neither makes up 99 percent of the atom's mass alone.

8. C: The total number of protons and neutrons in an atom is the atom's mass number. The number of protons in an atom is the atomic number. If an atom has a variation in the number of neutrons, the atom's mass number changes, but the atomic number remains the same. This variation creates isotopes of the atom with the same atomic number. The number of protons, Choice *A*, is unique for each atom and does not change. The number of orbitals and location of the nucleus remain the same for atoms; therefore, Choices *B* and *D* are incorrect. The number of electrons, Choice *E*, affect's the charge of the atom.

9. C: The rotational speed of all objects moving around an axis is the same. However, the linear speed of the object farthest away from the axis will be the greatest because they have a larger distance to cover in the same amount of time, compared with an object that is closest to the axis. On a merry-go-round, the innermost horse has the least distance to cover in a full rotation of the axis, so it will have the smallest linear speed; therefore, Choice *D* is incorrect. Horses anywhere in between the innermost and outermost horses will have linear speeds greater than that of the innermost horse but smaller than that of the outermost horse. Thus, Choices *A* and *B* are incorrect. The height of the horse would not affect the linear speed, so Choice *E* is incorrect.

10. D: The equation A + B → C represents a combination reaction because two reactants are being combined to form one larger product. Decomposition reactions, Choice *A*, are represented as C → B + A, where one larger reactant is broken into two smaller products. Combustion reactions, Choice *B*, always involve oxygen gas as a reactant. Although most reactions involve enzymes, this equation does not specify whether an enzyme is involved, so Choice *C* is incorrect. Oxidation/reduction reactions involve the oxidation of one species and the reduction of the other species via the transfer of oxygen, hydrogen, or electrons. Therefore, Choice *E* is incorrect.

11. B: The activation energy is the amount of energy that is required for a reaction to move forward and change the reactants into products. Catalysts decrease the amount of energy that needs to be generated to drive the reaction to happen, so they lower the activation energy of the reaction. The number of reactants and products, Choices *A* and *D*, remain the same in catalytic reactions, as do the concentration of reactants, so Choice *E* is also incorrect. Although increasing temperature can increase the rate of a reaction, it is not how catalysts work; thus, Choice *C* is incorrect.

12. E: The equation to determine torque is:
$$Torque = F_\perp \times distance\ of\ lever\ arm\ from\ the\ axis\ of\ rotation$$

where $F_\perp$ is the perpendicular force that is being applied to the object. Thus:

$$Torque = 35\ N \times 10\ m = 350\ N \times m$$

13. A: The molar mass of a substance is equivalent to the mass of one mole of the substance. It can be found on the periodic table for individual elements and is listed as the atomic mass unit. For example, the molar mass of carbon is 12.01 g/mol. One mole is equivalent to 6.02×10^{23} molecules of a substance, not only one molecule or ten molecules; therefore, Choices *B* and *E* are incorrect. The number of protons is the atomic number of an element, so Choice *C* is incorrect. The number of atoms is unrelated to the molar mass; thus, Choice *D* is incorrect.

14. B: Redox reactions are chemical reactions that involve one species of the reactants being oxidized and another species of the reactants being reduced. The transferring of hydrogen between reactants is one type of redox reaction. Another type involves the transferring of oxygen between reactants, and the third type involves the transferring of electrons between the reactants. Carbon, nitrogen, sulfur, and helium, Choices *A*, *C*, *D*, and *E* do not drive redox reactions.

15. B: Diffraction occurs when a wave travels through a slit or around an obstacle. The light bends and then causes a spread on the other side of the opening or obstacle. Refraction, Choice *A*, occurs when a wave travels from one medium to another and it bends as it moves through the different densities. Reflection, Choice *C*, occurs when a wave hits a medium that cannot pass through and is bounced back. The greenhouse effect, Choice *D*, occurs when light from the sun cannot escape the Earth's atmosphere and bounces back and forth by reflection. *Interference,* Choice *E*, occurs more generally with waves when an object affects the path of a wave or another wave interacts with that wave.

16. B: The elements are arranged such that the elements in columns have similar chemical properties, such as appearance and reactivity. Each element has a unique atomic number, not the same one, so Choice *A* is incorrect. Elements are arranged in rows, not columns, with similar electron valence configurations, so Choice *C* is incorrect. The names of the elements are not arranged alphabetically in columns, so Choice *D* is incorrect. Density is a physical property, not a chemical one, so elements in columns do not necessarily have the same density. Furthermore, the density of an element depends on the state it is in.

17. D: Refraction involves the bending of light through different mediums and does not affect the charge of an object. Electrons can be transferred from one object to another, giving an object a negative charge, so Choice *A* is incorrect. Induction, Choice *B*, involves the redistribution of charge by bringing a charged object close to two stationary objects. If the charged object is negatively charged, the electrons of the stationary objects will be repelled away and one of the stationary objects will take on a positive charge while the other one will become negatively charged. Polarization, Choice *C*, occurs when electrons in an object are reconfigured temporarily. Friction, Choice *E*, is a common way that objects are charged. The example of rubbing the socked feet along the carpet demonstrates how friction causes the transfer of electrons, and thus charge, from one object to another.

18. C: The pH scale goes from 0 to 14. A neutral solution falls right in the middle of the scale at 7. Choice *A*, a value of 13, indicates a strong base. Choice *B*, a value of 8, indicates a solution that is weakly basic. Choices *D* and *E*, 0 and 1, indicate strong acids.

19. A: Kinematics provides a guideline for describing motion. The equations describe the change in an object's position, Choice *D*, as a function of time, Choice *B*. Mass, Choice *C*, does not play a role in kinematics. Kinetics, not kinematics, involves forces; therefore, Choice *E* is incorrect.

20. D: Inertia is the idea of doing nothing and remaining unchanged. Newton's law of inertia describes what happens to an object when it stands alone, without outside forces causing changes in its motion. Choice *A* describes Newton's third law and involves action of an object. Choice *B* describes kinematics and motion. Choice *C* describes Newton's second law and involves a force acting on an object. Choice *E* describes momentum, not inertia.

Table Reading

The Table Reading section of the AFOQT assesses the test taker's ability to quickly and accurately read a table and locate the number indicated by the provided coordinates. The X-coordinates run horizontally along the top of the table labeling each column, and the Y-coordinates run vertically down the left side of the table labeling the rows. Typically, the range of values for X and Y run from a negative number in increments of 1 up to a positive number, for example from -8 to +8. For each of the 40 questions in the Table Reading section, an X-value and a Y-value are given, and test takers must quickly scan the table for the number located in the box indicated by the coordinates, which will be at their intersection. That number must then be selected among the five provided answer choices.

Although the task involved in finding the solutions to the Table Reading problems is quite simple from an intellectual demand standpoint, the section is quite difficult because of the sheer number of questions and the very limited amount of time. Speed is of the essence. In fact, there are 40 questions to complete in the allotted 7 minutes, which equates to just 10.5 seconds per question. As such, not only are speed and accuracy measured, but the setup provides the U.S. Air Force Admissions Officers with an opportunity to get a window into the test taker's ability to focus and stay calm under pressure and maintain composure. These are skills and qualities paramount to success as an Air Force candidate and future pilot.

Because there are so many numbers that inundate the test taker during this section—between the coordinates, the values that fill the table, and the potential answer options—it's extremely common to get flustered and forget what coordinates are sought, or worse, mix them up. For this reason, focusing first on just the X-coordinate, finding it, and locking in on that column while whispering the Y-coordinate to oneself can help retain the needed value while the correct X-row is followed down to the intersection of the intended Y-value. Physically using one's finger or a piece of paper to trace down the appropriate X-column will help prevent the test taker from inadvertently shifting columns while trying to travel down and meet up with the Y-value from its given row. Then, as soon as the desired value is attained in the box at the intersection of the coordinates, the test taker should quickly shift his or her glance to the answer choices, spot the matching value, and mark the letter corresponding to the correct answer.

Although speed is highly important in this section, it's actually detrimental for test takers to keep checking the clock. This will not only squander precious seconds, but it takes the focus off of the table reading and introduces yet more numbers and stress to the situation. The goal should be to stay focused on the table while working through the problems as efficiently and accurately as possible. When one's gaze is removed from the table to consider the time, the cognitive demand imposed by needing to refocus the attention on scanning the table values can mentally exhaust the brain and cause premature fatigue. This, in turn, will slow the subsequent processing of the table information. With that said, when there about 25-30 seconds remaining for the section, test takers should take note of where they are in their progress through the 40 questions. They should finish up the current question, and then simply fill in answers for the remaining problems. Because there is no penalty incurred for an incorrect answer, test takers should *always* mark an answer for each question, even if they have not had time to work through and solve the problem. Of course, the goal is always to fully address as many questions as possible on the Table Reading (and every other) section of the AFOQT, but when the time for a section has nearly elapsed, all additional questions should be marked with an answer choice.

While the goal of the practice questions for the Table Reading task is the familiarize the test taker with the task and practice the skills necessary to read the tables, it is important to set a timer to help replicate the speed challenge posed by this section.

Practice Questions

X value

	-4	-3	-2	-1	0	1	2	3	4
-4	22	23	24	26	27	28	30	32	33
-3	23	24	26	27	28	30	31	33	34
-2	24	25	27	28	30	31	33	34	35
-1	26	27	28	29	31	33	34	35	36
0	27	28	30	31	32	34	35	36	37
1	28	29	30	32	34	35	36	37	38
2	29	30	31	33	34	36	38	39	40
3	30	31	33	34	35	37	38	40	41
4	31	32	34	35	37	39	40	42	43

Y value

1. (-2, 1)
   a. 28
   b. 30
   c. 31
   d. 32
   e. 33

2. (1, 3)
   a. 35
   b. 39
   c. 37
   d. 40
   e. 34

3. (-4, 3)
   a. 32
   b. 33
   c. 29
   d. 35
   e. 30

4. (2, 2)
   a. 39
   b. 40
   c. 36
   d. 37
   e. 38

5. (3, 1)
   a. 35
   b. 38
   c. 40
   d. 37
   e. 36

6. (-4, -2)
   a. 24
   b. 30
   c. 26
   d. 28
   e. 25

7. (0, 3)
   a. 35
   b. 37
   c. 40
   d. 36
   e. 39

8. (1, -3)
   a. 28
   b. 29
   c. 33
   d. 30
   e. 31

9. (2, -4)
   a. 30
   b. 29
   c. 28
   d. 32
   e. 33

10. (-3, 0)
   a. 25
   b. 26
   c. 28
   d. 29
   e. 30

X value

	-4	-3	-2	-1	0	1	2	3	4
-4	33	34	35	37	38	39	41	42	43
-3	35	37	38	40	42	44	45	46	47
-2	36	38	39	41	42	44	46	47	48
-1	37	39	41	42	43	45	47	49	50
0	39	40	42	43	44	46	48	50	51
1	41	42	44	45	47	49	50	51	52
2	42	44	45	47	48	50	51	52	53
3	44	45	47	48	49	51	52	53	54
4	45	47	48	50	51	52	53	55	56

Y value

11. (-4, 1)
   a. 41
   b. 39
   c. 43
   d. 42
   e. 44

12. (-2, -1)
   a. 45
   b. 42
   c. 41
   d. 47
   e. 46

13. (3, 4)
   a. 50
   b. 56
   c. 53
   d. 55
   e. 54

14. (0, -4)
   a. 38
   b. 39
   c. 40
   d. 36
   e. 35

15. (3, 3)
   a. 52
   b. 53
   c. 55
   d. 56
   e. 58

16. (1, -4)
   a. 34
   b. 36
   c. 33
   d. 37
   e. 39

17. (-4, 0)
   a. 40
   b. 42
   c. 43
   d. 39
   e. 38

18. (-2, -3)
   a. 34
   b. 35
   c. 38
   d. 37
   e. 39

19. (-1, 2)
   a. 42
   b. 48
   c. 44
   d. 47
   e. 45

20. (3, -1)
   a. 50
   b. 49
   c. 53
   d. 48
   e. 54

		X value								
		-4	-3	-2	-1	0	1	2	3	4
	4	23	24	25	27	28	29	31	32	33
	3	24	26	27	28	29	31	32	33	34
	2	26	28	29	31	32	33	34	36	37
	1	27	29	31	32	33	34	36	38	39
Y value	0	29	30	32	33	34	36	37	39	40
	-1	30	31	33	35	37	38	39	41	42
	-2	31	33	34	36	37	39	41	42	43
	-3	32	34	36	37	39	40	42	43	44
	-4	34	36	38	39	41	43	44	45	46

21. (3, -1)
    a. 40
    b. 41
    c. 39
    d. 38
    e. 36

22. (4, 2)
    a. 31
    b. 33
    c. 35
    d. 37
    e. 38

23. (-2, 3)
    a. 30
    b. 32
    c. 29
    d. 28
    e. 27

24. (-1, -4)
    a. 39
    b. 40
    c. 42
    d. 38
    e. 37

25. (0, -1)
    a. 38
    b. 40
    c. 37
    d. 41
    e. 39

26. (3, 4)
    a. 29
    b. 32
    c. 31
    d. 33
    e. 27

27. (-1, -3)
    a. 33
    b. 34
    c. 37
    d. 32
    e. 36

28. (-2, 0)
    a. 32
    b. 33
    c. 35
    d. 38
    e. 36

29. (-3, 1)
    a. 30
    b. 27
    c. 26
    d. 29
    e. 25

30. (-1, 1)
    a. 33
    b. 34
    c. 30
    d. 29
    e. 32

X value

Y value	-4	-3	-2	-1	0	1	2	3	4
4	32	33	34	36	38	39	40	41	42
3	33	34	36	37	39	40	42	43	44
2	34	35	37	38	39	41	43	44	45
1	36	38	39	40	41	43	44	45	46
0	37	39	41	42	44	45	46	47	48
-1	38	40	42	43	45	46	48	50	51
-2	39	41	43	44	46	48	49	51	52
-3	41	43	44	46	48	50	51	52	53
-4	42	44	45	47	49	51	52	53	54

31. (-4, 2)
    a. 37
    b. 38
    c. 36
    d. 34
    e. 35

32. (-1, -1)
    a. 38
    b. 43
    c. 37
    d. 42
    e. 41

33. (3, 0)
    a. 42
    b. 41
    c. 44
    d. 46
    e. 47

34. (4, -2)
    a. 52
    b. 49
    c. 53
    d. 50
    e. 54

239

35. (-3, 4)
    a. 34
    b. 33
    c. 35
    d. 32
    e. 36

36. (-2, -3)
    a. 40
    b. 39
    c. 42
    d. 44
    e. 43

37. (3, 3)
    a. 44
    b. 47
    c. 45
    d. 42
    e. 43

38. (0, -3)
    a. 50
    b. 51
    c. 49
    d. 48
    e. 47

39. (1, -3)
    a. 50
    b. 52
    c. 53
    d. 54
    e. 56

40. (2, 0)
    a. 39
    b. 41
    c. 44
    d. 45
    e. 46

Answers

1. B	21. B
2. C	22. D
3. E	23. E
4. E	24. A
5. D	25. C
6. A	26. B
7. A	27. C
8. D	28. A
9. A	29. D
10. C	30. E
11. A	31. D
12. C	32. B
13. D	33. E
14. A	34. A
15. B	35. B
16. E	36. D
17. D	37. E
18. C	38. D
19. D	39. A
20. B	40. E

Instrument Comprehension

The Instrument Comprehension section is one of the few sections on the AFOQT where test takers can demonstrate their ability to understand the basic instrumentation used in aviation. Each problem will present a compass and an artificial horizon. Then, the answer choices present four planes in flight. Test takers must reconcile the position information indicated by the instruments, which in aggregate, give the airplane's compass heading, degree of banking, and amount of climb or dive. They must then select the airplane silhouette that most closely models this position. The test taker's viewpoint is always looking north and is shown from the same altitude as each of the four silhouetted airplanes. The planes may be shown from a front view, rear view, or side view, and may be banking, climbing, diving, or level. As test takers look at the test page, east is located to the right of the page.

Usually, the circle furthest to the left in each question is labeled as the artificial horizon. In the center of the circle, there is a small, stationary aircraft silhouette. There is a heavy black line, which represents the line of the horizon, and a black pointer that indicates the aircraft's degree of banking to the left or right. The positions of these two indicators vary according to the position of the airplane in which the instrument on which they are contained is located.

When the airplane is neither banking to the left nor to the right, the black pointer will be directed toward zero. This means that the plane is not performing a turning maneuver; it is flying straight forward. The pointer directed toward the left of zero indicates that the airplane is banking to the pilot's right, while the pointer directed toward the right of zero indicates that the airplane is banking to the pilot's left. The heavy black line of the horizon will be tilted when the aircraft is banking, but it is always oriented 90-degrees from the pointer.

When the airplane is level and neither diving nor climbing, the horizon line will be directly on the silhouetted airplane's fuselage. This means that the nose of the airplane is level with the ground. When the airplane is climbing, the silhouette will be located between the pointer and the horizon line. The distance between the silhouetted airplane and the horizon line varies according to the degree of climb, so that the further the distance between the silhouette and the horizon line, the more significant the climb. The horizon line located between the pointer and the silhouette indicates that the airplane is diving. Again, the further apart that the horizon line and the silhouette, the more substantial the dive.

Typically, the compass appears on the right side of the question and shows the direction in which the airplane is headed. For example, if the needle on the compass is pointing at the "W," the airplane is headed west. Sometimes, the needle may between one of the four cardinal directions. If it is pointing between the "E" and the "S," for example, the airplane is heading southeast. This is exactly how a normal compass used for orientation operates.

On the exam, the directions work like this:

North: Aircraft flying away from you (rear of plain visible)
South: Aircraft flying towards you (front of plain visible)
East: Aircraft nose to the right of the page
West: Aircraft nose to the left of the page

For the purpose of these questions, the simulated aircraft will appear to be flying away from your view for north, facing you for south, facing with the nose to left of the page for west and to the right for east.

Test takers will likely be familiar with this instrument or they can attain additional practice by downloading a compass application or purchasing a small compass from a department store or sporting goods retailer.

Test takers are given five minutes to work through 25 Instrument Comprehension problems, which equates to just under 12 seconds per question. Of all the sections of the AFOQT, Instrument Comprehension often warrants additional practice time because it may be the least familiar to test takers. It also most closely resembles skills directly employed by Air Force Officers, so it is important for candidates to demonstrate command and high achievement on this section to impress admissions officers. It is critically important for test takers to familiarize themselves with reading the compass and artificial horizon and be able to translate the information they convey to the position of the airplane so that the correct silhouette can be identified.

As with all sections on the AFOQT, there is no penalty incurred for incorrect answers. As such, test takers should *always* mark an answer for each question, even if they have not had time to work through and solve the problem. When the time remaining in the section is limited to 20-30 seconds, the unaddressed questions should each be marked with an answer.

Practice Questions

1.

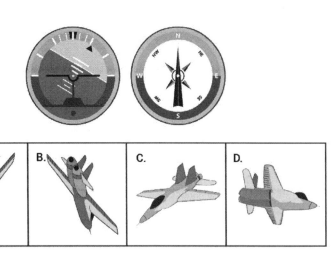

A.	B.	C.	D.

2.

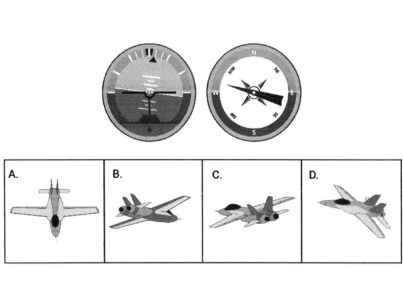

A.	B.	C.	D.

3.

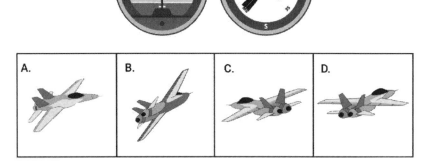

A.	B.	C.	D.

4.

A.	B.	C.	D.
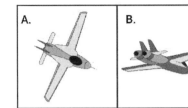		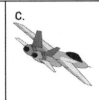	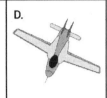

5.

A.	B.	C.	D.
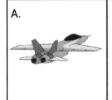			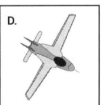

6.

A.	B.	C.	D.

7.

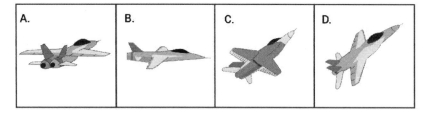

A.	B.	C.	D.

8.

A.	B.	C.	D.

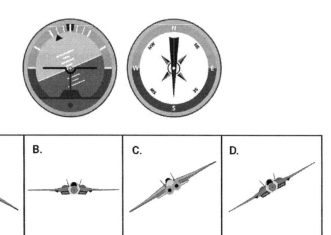

9.

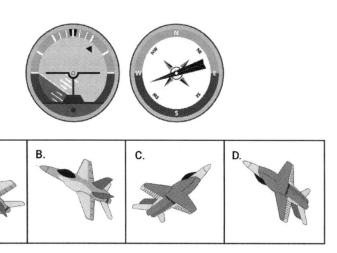

A.	B.	C.	D.

10.

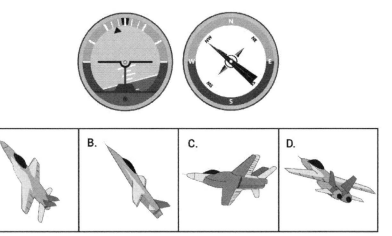

11.

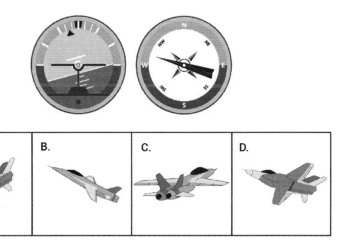

12.

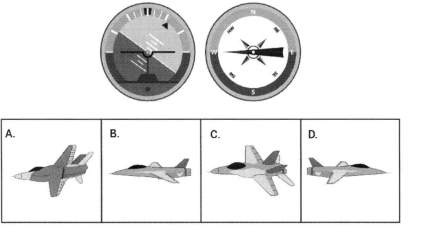

247

13.

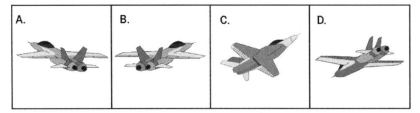

14.

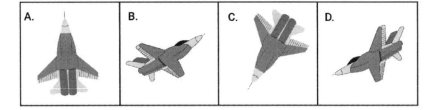

15.

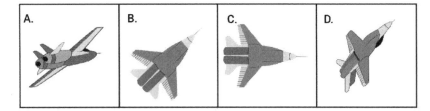

16.

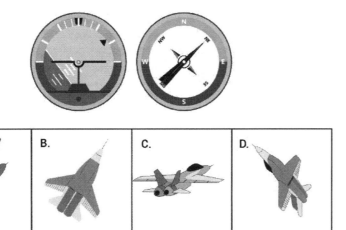

A.	B.	C.	D.

17.

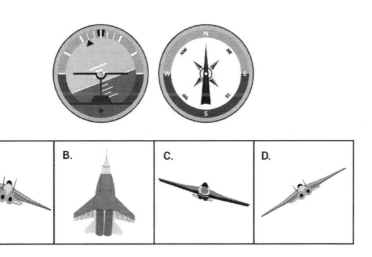

A.	B.	C.	D.

18.

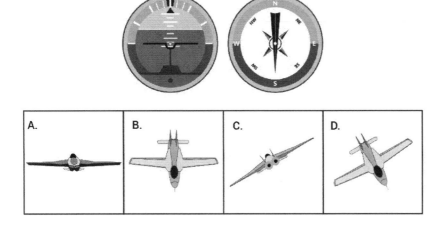

A.	B.	C.	D.

249

19.

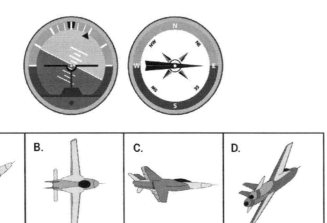

A.	B.	C.	D.

20.

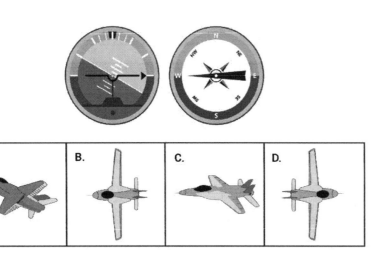

A.	B.	C.	D.

21.

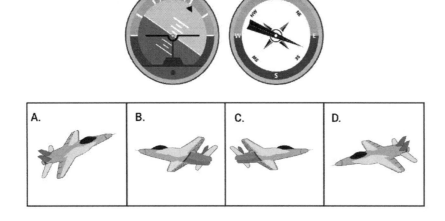

A.	B.	C.	D.

22.

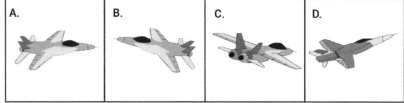

A.	B.	C.	D.

23.

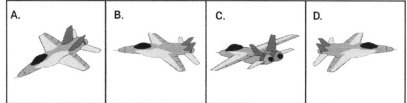

A.	B.	C.	D.

24.

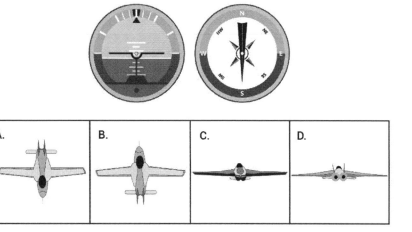

A.	B.	C.	D.

25.

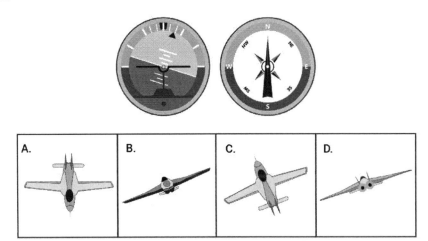

Answers

1. A

2. C

3. D

4. A

5. C

6. B

7. C

8. D

9. B

10. D

11. D

12. C

13. B

14. D

15. A

16. A

17. A

18. B

19. C

20. C

21. C

22. C

23. A

24. C

25. D

Block Counting

The Block Counting section of the AFOQT measures the test taker's spatial and logical reasoning skills. These skills are assessed by evaluating the test taker's ability to fully visualize a three-dimensional pile of blocks presented in a two-dimensional image. The two-dimensional rendering will have certain faces of blocks hidden by the loss of information that occurs when a three-dimensional structure is reduced to a two-dimensional form. In the image, certain blocks will be numbered, and the test taker must focus on the block addressed in the question and determine the number of other blocks that specific labeled block touches. In this way, the Block Counting exercise measures test takers' deductive reasoning ability by requiring them to envision the arrangement of blocks and their faces that aren't necessarily explicitly shown because they extend into a concealed side or section of the pile.

There are 30 Block Counting questions to address in the allotted 4.5 minutes, which equates to a mere 9 seconds per question. There are four different stacks of blocks presented in this section—each followed by five questions that pertain to the same stack but address different blocks in the stack. Each question has five answer choices.

All of the blocks stacked in each pile are of the exact same size and shape, so test takers can be confident that there are not differently sized blocks in each stack. In the practice questions below, the blocks may not be perfectly to scale but they are all considered the exact same size. This makes it easier to deduce how the blocks must extend and touch one another, even when considering the inside or backside of the pile. However, it should be noted that the orientations of the blocks in a stack will vary. Blocks, which are rectangular prisms, may be oriented vertically with the long side going up and down, horizontally, or on their side. This adds to the challenge of visualizing the stack in three-dimensions and inferring the number of blocks to count that must touch the block of interest.

Although not all of the blocks nor all of their faces will be visible in the image, using logic and spatial reasoning skills can help the test taker fill in the "missing pieces." When trying to determine if a particular block "touches" another one, test takers must only consider if the faces of the blocks touch one another. Corner-only contact does not qualify as touching interaction to be counted for a given block.

The speed required to answer all 20 questions in this section adds a significant layer of difficulty. Test takers also need to keep straight in their minds which blocks they have already accounted for in instances where multiple faces of one block may abut the block in question. It is important for test takers to imagine how the blocks must connect and not allow the time crunch imposed by the section to cause careless mistakes, either by mistakenly counting certain connections twice or missing others all together. Each rectangular prism, by nature, has six faces. It is possible though for a block to have multiple blocks touching a given face because of the potential orientation of shorter, thinner, or narrower sides touching a broader face of a block.

As with all sections on the AFOQT, because there is no penalty incurred for incorrect answers, test takers should *always* mark an answer for each question, even if they have not had time to work through and solve the problem. Of course, the goal is always to fully address as many questions as possible on the Block Counting (and every other) section of the AFOQT, but when the time for a section has nearly elapsed, all additional questions should be marked with an answer choice.

While the goal of the practice questions for the Block Counting task is to familiarize the test taker with the task and practice spatial reasoning skills, it is important to set a timer to help mimic the significant speed challenge posed by this section.

Practice Questions

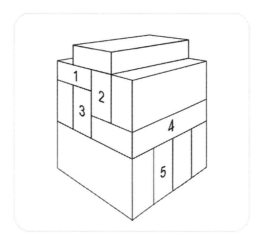

1. Block 1 is touched by ___ other blocks.
 - a. 3
 - b. 4
 - c. 5
 - d. 6
 - e. 7

2. Block 2 is touched by ___ other blocks.
 - a. 3
 - b. 4
 - c. 5
 - d. 6
 - e. 7

3. Block 3 is touched by ___ other blocks.
 - a. 4
 - b. 5
 - c. 6
 - d. 7
 - e. 8

4. Block 4 is touched by ___ other blocks.
 - a. 4
 - b. 5
 - c. 6
 - d. 7
 - e. 8

5. Block 5 is touched by ___ other blocks.
 - a. 4
 - b. 5
 - c. 6
 - d. 7
 - e. 8

6. Block 1 is touched by ___ other blocks.
    a. 5
    b. 6
    c. 7
    d. 8
    e. 9

7. Block 2 is touched by ___ other blocks.
    a. 3
    b. 4
    c. 5
    d. 6
    e. 7

8. Block 3 is touched by ___ other blocks.
    a. 5
    b. 6
    c. 7
    d. 8
    e. 9

9. Block 4 is touched by ___ other blocks.
    a. 5
    b. 6
    c. 7
    d. 8
    e. 9

10. Block 5 is touched by ___ other blocks.
    a. 5
    b. 6
    c. 7
    d. 8
    e. 9

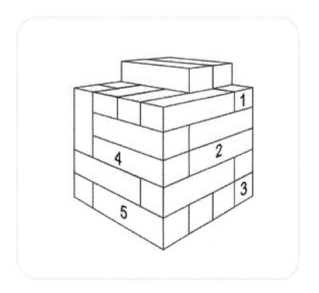

11. Block 1 is touched by ___ other blocks.
    a. 5
    b. 6
    c. 7
    d. 8
    e. 9

12. Block 2 is touched by ___ other blocks.
    a. 3
    b. 4
    c. 5
    d. 6
    e. 7

13. Block 3 is touched by ___ other blocks.
    a. 3
    b. 4
    c. 5
    d. 6
    e. 7

14. Block 4 is touched by ___ other blocks.
    a. 5
    b. 6
    c. 7
    d. 8
    e. 9

15. Block 5 is touched by ___ other blocks.
    a. 3
    b. 4
    c. 5
    d. 6
    e. 7

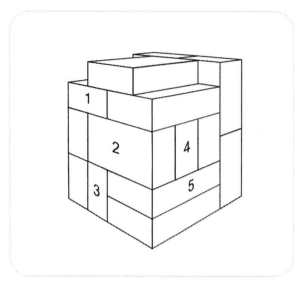

16. Block 1 is touched by ___ other blocks.
    a. 5
    b. 6
    c. 7
    d. 8
    e. 9

17. Block 2 is touched by ___ other blocks.
    a. 3
    b. 4
    c. 5
    d. 6
    e. 7

18. Block 3 is touched by ___ other blocks.
    a. 3
    b. 4
    c. 5
    d. 6
    e. 7

19. Block 4 is touched by ___ other blocks.
    a. 5
    b. 6
    c. 7
    d. 8
    e. 9

20. Block 5 is touched by ___ other blocks.
    a. 5
    b. 6
    c. 7
    d. 8
    e. 9

Answer Explanations

1. B: This first set of blocks was the easiest of the four. There are no hidden blocks. Block 1 has 1 block on top of it, 2 blocks on bottom, 0 blocks to the left and 1 block to the right. This gives a total of 4 blocks.

2. C: Block 2 has 1 block on top of it, 1 block on the bottom, 2 blocks to the left, and 1 block to the right. This gives a total of 5 blocks.

3. E: Block 3 has 1 block on top of it, 4 blocks on the bottom, 1 block on the left, and 2 blocks on the right. This gives a total of 8 blocks.

4. D: Block 4 has 2 blocks on top of it, 4 on the bottom, 0 to the left, 0 to the right, and 1 block behind it. This gives a total of 7 blocks.

5. B: Block 5 has 3 blocks on top of it, 0 blocks on the bottom, 1 block to the left, and 1 block to the right. This gives a total of 5 blocks.

6. D: For this set of blocks, it is easiest to think of each block as a 2x1. If the blocks were made into cubes, then each of these blocks would be 2x1. With this being the case, we can know that no block will be touched by more than 2 blocks on any long side and can only be touched by one block on its short side. So, block one has 2 touching it on top, 2 on the left, 1 on the right, 2 on bottom, and 1 on the back side. This gives a total of 8 blocks.

7. E: Block 2 has 2 blocks touching it on top, 1 on the left, 1 on the right, 2 on the bottom and 1 on the back. The back side can be determined to only have one block touching it by considering the layout of block 4. This means there has to be another block behind block 2 in the exact same orientation as block 2. Otherwise, there would be a void in the middle of this stack.

8. A: Block 3 has 2 blocks on top, 1 on the left, 1 on the right, 0 on the bottom, and 1 on the back. This gives a total of 5 blocks.

9. E: Block 4 has 2 blocks touching it on top, 2 blocks on the left, 2 blocks on the right, 2 blocks on bottom, and 1 block on the back. This gives a total of 9 blocks.

10. C: Block 5 has 2 blocks touching it on top, 0 blocks on the left, 1 block on the right, 2 on the bottom, and 2 on the back. This gives a total of 7 blocks.

11. D: This set of blocks is similar to the last set except the blocks are 3x1. So, block 1 has 1 block touching it on top, 3 on the left, 0 on the right, 3 on the bottom, and 1 on the back. This gives a total of 8 blocks.

12. C: Block 2 has 1 block touching it on top, 2 blocks on the bottom, 1 block on the left, 0 on the right, and 1 in the back. This gives a total of 5 blocks.

13. A: Block 3 has 1 block touching it on top, 0 blocks on the bottom, 1 block on the left, 0 blocks on the right, and 1 block in the back. This gives a total of 3 blocks.

14. C: Block 4 has 1 block touching it on top, 2 on the bottom, 1 on the right, 0 on the left, and 3 in the back. This gives a total of 7 blocks.

15. B: Block 5 has 2 blocks touching it on top, 0 blocks on the bottom, 1 block on the left, 0 blocks on the right, and 1 block in the back. This gives a total of 4 blocks.

16. C: In this set of blocks, every block is a 1x2x3. It is a single-block wide, while its height is double, and its length is three times. Block 1 has 1 block touching it on top, 0 blocks on the left, 1 block on the right, 4 blocks on bottom and 1 block in the back. This is a total of 7 blocks.

17. D: Block 2 has 2 blocks touching it on top, 2 blocks on the bottom, 1 block on the left, 0 blocks on the right, and 1 block in the back. This gives a total of 6 blocks.

18. E: Block 3 has 3 blocks touching it on top, 0 blocks on the bottom, 1 block on the left, 2 blocks on the right, and 1 block touching it in the back. This gives a total of 7 blocks.

19. C: Block 4 has 2 blocks touching it on top, 2 blocks on the bottom, 1 block on the left, 1 block on the right, and 1 block in the back. This gives a total of 7 blocks.

20. B: Block 5 has 3 blocks touching it on top, 1 block on the bottom, 0 blocks on the left, 1 block on the right, and 1 block in the back. This gives a total of 6 blocks.

Aviation Information

Basic Flight Concepts and Terms

Aerodynamics Terms

Aerodynamics: Aerodynamics is the study of the motion of air as it encounters a solid object such as an airplane wing. Balancing the four aerodynamic forces (lift, drag, weight, and thrust) to overcome the force of gravity makes the flight of heavier-than-air objects possible.

Lift: Lift causes the object to remain in the air. Lift is the force applied perpendicular to the direction of the object's motion to exceed the force of gravity. Lift is created by airflow over an airfoil surface such as the wing of an aircraft. When air flows faster over the airfoil surface than underneath, lift is achieved.

Drag: Drag resists the airflow that enables motion. Drag is the force applied parallel to the direction of the object's motion. Drag must be countered by an equal or greater thrust in order to achieve or maintain motion.

Weight: Weight is the force that draws an object to Earth. It is the combination of the mass of the object and the force of gravity. Weight limits flight, as gravity exerts constant pressure to draw the object down to Earth. The thrust must be sufficient to propel the object into the air.

Thrust: Thrust or power is the force that propels the aircraft into the air. Thrust is produced by accelerating the mass of air around the aircraft by one of the means of propulsion. Thrust is resisted by drag.

Airfoil: An airfoil is a surface such as an aircraft wing or rotor blade that is shaped to split airflow above and below the airfoil to create lift. As the air over the top of the airfoil speeds up, the pressure above the airfoil decreases. At the same time, the pressure below the airfoil surface increases and creates lift.

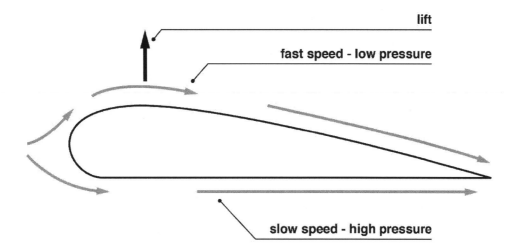

Angle of Attack: The angle of attack is the angle at which the airfoil encounters incoming air. The angle of attack is measured as the angle between the chord line of the airfoil surface and the motion of the airflow in relation to the airfoil.

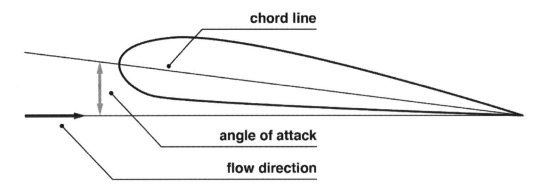

Increasing the angle of attack increases the *lift coefficient*. If the angle of attack exceeds the maximum value of the lift coefficient, the separation of the airflow over the airfoil surface becomes too great, and the lift coefficient begins to decrease.

Lift Coefficient: The lift coefficient is the factor used to calculate the lift of the airfoil. The number takes into account the airspeed of the aircraft, the density of the air around the airfoil, the area of the airfoil surface or wing, and the angle of attack.

Chord Line: The chord line is the imaginary line from the front to the back of the airfoil surface (leading and trailing edges) as in the illustration above. The chord length is the distance of this line from the leading edge to the trailing edge.

Critical Angle of Attack: The critical angle of attack is the angle of attack that produces the maximum value of the lift coefficient. As the angle increases below the critical angle of attack, the lift coefficient increases. If the angle exceeds the critical angle of attack, the lift coefficient decreases.

Types of Motion

The types of motion of the aircraft around its center of gravity are used to control the aircraft and maintain stability and balance.

Yaw: Yaw is the rotation of the aircraft around its vertical axis. Yaw affects the direction in which the aircraft is heading. It is controlled by changing the direction that the nose of the aircraft is pointed. Yaw is used in combination with roll to turn the aircraft.

Roll: Roll is the rotation of the aircraft around its longitudinal (front to back) axis. When lift is decreased on one wing and increased on the other, the aircraft will roll left or right in the direction in which lift is decreased. Pilots use **ailerons**—flight control surfaces on the wings—to control the direction of the roll. This is called **banking**.

Pitch: Pitch is the rotation of the aircraft around its lateral (side to side) axis. To control the pitch, the pilot tilts the nose of the aircraft up or down. When the nose is tilted up, the aircraft will climb. If it is tilted down, the aircraft will descend. When the pitch is level, the aircraft remains level in relation to the horizon.

Stall: Stall occurs when a sudden decrease in lift and increase in drag causes the aircraft to fall. Lowering the angle of attack corrects a stall.

Spin: A spin occurs as a result of a stall that is improperly corrected. When an aircraft enters a flat spin, the pilot must disrupt the spin to force the craft back into a stall.

x.   Phases of Spin:
- o   Incipient: Phase in which the aircraft has begun to spin after a stall.
- o   Fully Developed: Phase in which the aircraft is descending in a near-vertical spiral.
- o   Flat Spin: Phase in which the aircraft is spinning around its own center of gravity.

Airspeed

Airspeed is the speed of the aircraft measured against the speed of sound (Mach 1). A sonic boom is the sound emitted when an aircraft reaches Mach 1.

y.   Types of speed:
- o   Subsonic: Less than Mach 1.
- o   Transonic: Near Mach 1
- o   Supersonic: Over Mach 1
- o   High Supersonic: Mach 3–Mach 4 (3-5x the speed of sound)
- o   Hypersonic: Mach 5–10 (5–10x the speed of sound)
- o   High Hypersonic: Higher than Mach 10.

Means of Propulsion

In order to provide thrust, an aircraft requires a means of propulsion, which differs depending on the design of the aircraft.

Propeller Engine: A system in which the engine mixes fuel with air and burns the fuel to release heated gas that moves a piston attached to a crankshaft to spin a propeller.

Rocket Engine: A system in which the engine combines fuel and oxidizers in a combustion chamber to release hot exhaust.

Gas Turbine Engine (Jet Engine): A system in which the engine combines fuel and air from the surrounding atmosphere in a combustion chamber to release hot exhaust.

Fixed-Wing Aircraft Structure

A **fixed-wing aircraft** is an aircraft with stationary wings attached to a fuselage that uses forward airspeed to generate lift. An airplane is a fixed-wing aircraft. Fixed-wing aircrafts can be constructed out of wood, aluminum, carbon fiber, and other materials.

A fixed-wing aircraft consists of the following components:

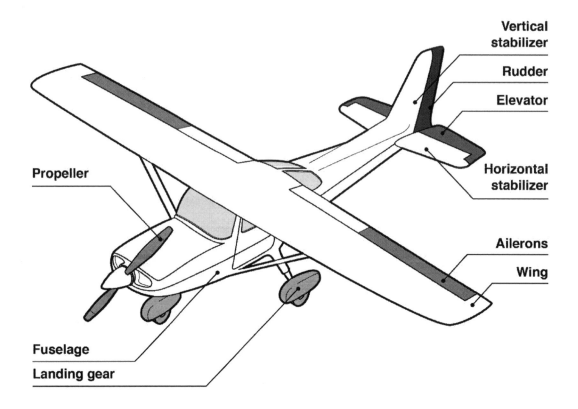

Fuselage

The **fuselage** is the main supporting structure of the aircraft that carries the pilot, passengers, and cargo. The fuselage includes the cockpit that contains the controls of the aircraft. The engine may be contained within or affixed to the fuselage in a nacelle.

The fuselage, along with the other components of a fixed-wing aircraft, is constructed to withstand the stress imposed upon the aircraft. The types of stress for which the aircraft design must account include the following:

z. Bending: A force applied perpendicular to an object that causes an object to bend.
aa. Compression: A force that pushes an object and reduces its size.
bb. Shear: A force applied parallel to an object that causes layers of an object to shift or break off.
cc. Tension: A force that pulls an object and causes the object to stretch.
dd. Torsion: A force that applies torque that twists an object.

Fuselage Construction
A fuselage may be constructed from one of the following types:

Truss: A rigid frame that includes bars, beams, and struts designed to withstand tension and compression. Truss construction includes **longerons** (also called **stringers** or **stiffeners**) that provide length-wise (longitudinal) support of the frame and diagonal bracing.

Monocoque: A single-shell frame that relies on the strength of its skin or covering to withstand stress. Monocoque frame assembly uses formers that determine the shape of the fuselage and bulkheads to resist pressure.

Semimonocoque: A hybrid frame that consists of both a strong outer skin, formers, and bulkheads like a monocoque frame, as well as longerons and diagonal bracing to support the structure like a truss frame.

Wings

The wings of a fixed-wing aircraft are the airfoil surfaces that generate lift.

Wing Shape
The shape and size of the leading edge—the front edge of the airfoil section—and the trailing edge—the rear edge of the airfoil section—affect the lift, balance, and stability of an aircraft and are determined by the desired flight conditions. Some common variations include:

ee. Straight leading and trailing edges
ff. Tapered leading and trailing edges
gg. Straight leading edge and tapered trailing edge
hh. Tapered leading edge and straight trailing edge

Wing Configuration
Fixed-wing aircrafts have a pair of wings arranged symmetrically to the left and right of the fuselage. This pair of wings is called the **wing plane**. A fixed-width aircraft may have one or multiple wing planes. A biplane is an aircraft with two wing planes; a triplane has three wing planes; a quadruplane has four wing planes; and a multiplane has many wing planes.

The wings are attached to the top, bottom, or middle of the fuselage of the aircraft. The location and the angle of the wings along the horizontal plane of the fuselage affect the lateral stability and performance of the aircraft. Some common wing attachment locations and angles include the following:

ii. Low wing: straight and horizontal, attached to the bottom of the fuselage
jj. Mid wing: straight and horizontal, attached to the middle of the fuselage
kk. High wing: straight and horizontal, attached to the top of the fuselage
ll. Dihedral: angled up or down, attached to the top or bottom of the fuselage
mm. Gull wing: bent upward toward the wing root, attached to the top or bottom of the fuselage
nn. Inverted gull: bent downward near the wing root, attached to the top or bottom of the fuselage

Wing Structure

A wing must support itself in order to lift the aircraft into the air. Adding external bracing can reduce the weight of a heavy wing but might cause additional drag that can affect the performance of the aircraft. Most modern aircrafts use a cantilevered design that reduces drag and supports itself with the use of internal structure and aerodynamic skin. Other aircrafts use a braced design, an external support structure that uses rigid struts that act in compression and tension, or wires that act in tension. A braced design often uses both struts and wires to support the wing. An aircraft may also use a combination of cantilevered and braced wing designs.

Wing Construction

Wing spars provide the main structure of the wing. Wing spars support the weight of the fuselage and engines, as well as the load the aircraft is carrying. Like the longerons of the fuselage, wing spars run span-wise from the fuselage (wing root) to the tip of the wing. Wings may have one or multiple spars. In addition to wing spars, wing ribs, formers, and/or bulkheads cross the spars from the leading end to trailing end of the wing to provide the shape of the wing and help carry the load and stress caused by flight. The number and arrangement of wing spars and wing ribs can vary depending on the required flight conditions.

The **wing skin** provides additional support. A rigid stressed-skin design relieves some of the load and stress of the flight from the spars. The panels that form the wing skin can be made from a variety of materials including aluminum and wood, as well as fiberglass and carbon fiber.

Flight Control Surfaces

Flight control surfaces are the hinged flaps on an aircraft that are used to manipulate the lateral, vertical, and longitudinal directions to steer the aircraft through the sky. The primary flight control surfaces used for aerial navigation include the elevator, rudder, and ailerons. As previously discussed, these flight surfaces control the pitch (lateral), yaw (vertical), and roll (longitudinal), respectively.

The primary flight control surfaces include the following:

Rudders

Flight control surface attached to the vertical stabilizer (tail wing) that controls the yaw and counters the adverse yaw created by the ailerons that causes the aircraft to yaw in the opposite direction of the roll.

Ailerons

Flight control surfaces attached to the wings on the trailing edges that control the roll and, subsequently, the lateral balance. The aileron goes up on the side of the roll. For example, when the aircraft rolls right, the right aileron goes up. The opposite aileron goes down. Lift and drag increase for the raised wing, which causes the aircraft to yaw in the direction of the raised wing, opposite the roll (i.e. adverse yaw).

Elevators

Flight control surfaces attached to the horizontal stabilizer (tail wing) that controls the pitch (up and down orientation of the nose, sometimes called the nose attitude) by changing the angle of attack on the horizontal stabilizer. The up and down motion of the aircraft can help control airspeed, as it increases or decreases the lift applied to the tail. Tilting the nose down will increase lift but will result in lost altitude, while tilting the nose up decreases the lift and speed, requiring thrust to be applied to increase altitude.

Booms

Booms are the rear fuselage of an aircraft affixed to the main wings on either side. Booms can contain fuel tanks, extend the tail, or provide additional support. Booms are typically used in a twin-boom configuration that allows the aircraft to have a large engine or access door at the back of the main fuselage.

Nacelles

Nacelles are compartments attached to the fuselage or built into the wings that contain the engine and engine components, the aircraft firewall, which separates the engine from the cockpit, and/or the landing gear of the aircraft. A nacelle is covered by a skin and features a cowling that allows access to the engine or equipment. The structure of a nacelle is similar to that of the frame assembly of a fuselage that includes longerons, diagonal bracing, and bulkheads.

Cowling

The **cowling** is the detachable panel that serves as the cover of the aircraft engine to facilitate access. The cowling on a nacelle can also serve as the panel that opens when the landing gear is retracted and lowered. Cowlings on an aircraft may also be used to allow airflow into the engine to cool it or to intake the air necessary for combustion in a gas turbine engine.

Fairings

Fairings are used on a fixed-width aircraft to seal the spaces between the various components, improve the appearance, and smooth sharp edges to reduce drag. Fairings are often used at the wing root and wing tip, rudder, on the elevator and aileron flaps, on stabilizers, around the cockpit, and on the tail cones, among others.

Landing Gear

Landing gear is deployed to support the plane when it lands. Aircrafts may have retractable landing gear, which is pulled up during the flight to prevent drag, or fixed landing gear that remains in place through the flight.

Although different types of landing gear can be used depending on the aircraft's purpose, typical landing gear consists of braces with wheels in the following arrangements:

Conventional undercarriage: Arrangement with two front wheels and one smaller wheel or skid in back that is used on older model aircrafts.

Tricycle undercarriage: Arrangement with one small wheel in front and two wheels in back that is used on modern aircrafts.

Four Fundamental Flight Maneuvers

Flight maneuvers are movements performed by the aircraft under the control of a pilot. Maneuvers are performed by adjusting the flight control surfaces to manipulate the pitch, roll, and yaw.

The flight control surfaces are controlled as follows:

To make the nose of the aircraft go up (pitch), apply back pressure to lower the elevator.

To make the nose of the aircraft go down (pitch), apply front pressure to raise the elevator.

To make the right or left wing go down (roll), apply right or left pressure to raise the aileron on the right or left wing, respectively.

To make the aircraft turn left or right on the vertical axis (yaw), apply pressure to the left or right rudder pedal, respectively.

Trimming refers to releasing pressure on the flight controls to maintain a consistent attitude and airspeed. Trimming is accomplished by engaging trim tabs located on the elevator, rudder, and/or ailerons.

Basic controlled flight relies on the application of the flight controls to perform the following four fundamental flight maneuvers: straight and level flight, turns, climbs, and descents.

Straight and Level Flight

Straight and level flight is controlled flight in which the heading (straight) and the altitude (level) remain steady. Straight flight is achieved by maintaining a level wing plane in relation to the horizon. To do this, the pilot must adjust for unintentional roll (turns) by applying pressure to the right and left ailerons as needed to maintain consistent heading. Level flight is achieved by maintaining a straight nose in relation to the horizon. For this, the pilot must adjust for unintentional pitch (climbs or descents) by applying pressure to the front and back of the elevator and trimming as needed.

Turns

To turn an aircraft, the pilot banks left or right to achieve roll in the direction they want to go. Banking requires applying pressure to raise the aileron to lower the wing. The aircraft turns in the direction of the bank.

The three types of turns include the following:

oo. Shallow: A turn in which the bank is less than 20 degrees.
pp. Medium: A turn in which the bank is approximately 20-45 degrees.
qq. Steep: A turn in which the bank is more than 45 degrees.

Climbs

A climb increases altitude. During a climb, the pilot applies pressure to lower the elevator, which raises the nose (pitch) of the aircraft. When an aircraft is in a climbing attitude, the upward direction of the weight of the aircraft causes the drag to increase. To counter the drag, the pilot must apply more thrust. An aircraft may only sustain a climbing attitude as long as enough thrust is available to counter the drag.

Descents

A descent decreases altitude. To descend, the pilot raises the elevator to angle the nose of the aircraft down. As weight shifts down and the nose is no longer parallel with the oncoming airflow, drag deceases, causing an increase in airspeed. To descend safely, the pilot must decrease the thrust while the forward motion of the aircraft allows gravity to draw it down to Earth. This descending maneuver is called a glide.

Helicopters

A helicopter is a rotary-wing aircraft that uses a rotor to lift off the ground and move through the air. The rotor is comprised of an array of small airfoils (rotor blades) that spin constantly around a shaft. Since the rotor blades of a helicopter are constantly moving, helicopters, unlike fixed-wing aircrafts, can stop and hover above the ground. Additionally, helicopters do not require a runway to land, so they can land in places airplanes cannot.

Parts of a Helicopter

Main Rotor Blade

The main rotor blade is the helicopter's airfoil and serves the same function as an airplane wing: to generate lift. Lift is generated by the blades as they spin horizontally around the rotor mast. The pilot can manipulate the rotor's rotations per minute (rpms) and its angle of attack to generate more or less lift, as needed. The main rotor also serves as the means to steer the helicopter. The pilot can manipulate the angle of the main rotor to control the motion of the helicopter.

Swash Plate Assembly

The swash plate assembly communicates the input from the flight controls to the rotor blades to control the motion of the helicopter. The swash plate assembly includes the upper swash plate and lower swash plate. The upper swash plate is connected to the rotor mast and rotates to communicate the input from the pilot through control rods that link the swash plate to the rotor blades. The upper swash plate receives the input from the pilot through the lower swash plate, which is linked through control rods to the cyclic and collective-pitch levers. The lower swash plate remains stationary, and ball bearings separate the swash plates, so the upper swash plate can rotate freely.

Stabilizer Bar

The stabilizer bar, or **flybar**, helps stabilize the helicopter. The bar is attached to the top of the main rotor and uses weights on either end to maintain a consistent rotation that decreases the vibrations from the forces acting on the rotor blades from the atmosphere and from the pilot manipulating the flight controls.

Rotor Mast

The rotor mast is the shaft around which the rotor blades spin. The rotor mast connects the rotor blades to the transmission of the helicopter.

Transmission

The transmission of a helicopter transmits the power from the engine to the rotors and controls the speed of the rotation of the main and tail rotor blades.

Engine

The engine of the helicopter is its means of propulsion. This is often a gas turbine engine that generates thrust powerful enough to lift the helicopter's weight off the ground and maintain the constant rotary motion required to keep the helicopter in the air.

Fuselage
Like a fixed-wing aircraft, helicopters have a fuselage that acts as its main supporting structure and contains the cockpit.

Cyclic and Collective Pitch Levers
A pilot steers a helicopter by adjusting the pitch using the cyclic and collective pitch levers. The cyclic pitch lever controls the side to side and the forward and backward motion of the helicopter by tilting the angle of the pitch. The collective pitch lever controls the up and down motion of the helicopter by increasing or decreasing the pitch.

Foot Pedals
The foot pedals control the direction of the helicopter by controlling the direction the tail rotor spins. The right pedal turns the nose of the helicopter to the right and the tail to the left while the left pedal turns the nose to the left and the tail to the right.

Tail Rotor
The tail rotor spins vertically in the opposite direction to the main rotor to control the direction of the helicopter. The tail rotor prevents the main rotor from spinning the fuselage in a circle around its center of gravity by compensating for the torque created by the main rotor. The torque generated by the tail and main rotor should balance each other. If the torque of the tail rotor is greater than that of the main rotor, the helicopter will yaw to the left. If it is less, the helicopter will yaw to the right.

Tail Boom
The tail boom is the extension of the fuselage on which the tail rotor and the vertical and horizontal stabilizers are mounted. It assists the tail rotor to counter the torque generated by the main rotor and stabilize the aircraft.

Cockpit
The cockpit of the helicopter houses the pilot and co-pilot and the flight controls. These include the cyclic and collective-pitch levers.

Landing Skids
The landing skids of a helicopter, like the landing gear of a fixed-wing aircraft, serve to support the helicopter as it lands. Skids are lighter and less expensive than wheels and are more suited to smaller helicopters that need to lift heavy loads or hover, while wheels are more suitable for larger helicopters. As landing skids do not significantly increase drag, they are not typically retractable.

Flight Envelope

The **flight envelope**, also called the **service** or **performance envelope**, refers to the range of minimum and maximum performance capabilities within which an aircraft can safely operate. The performance capability of an aircraft is measured in terms of the maximum potential speed versus its maximum potential load factor. Aircrafts must be able to operate within the specified flight envelope for its design to be certified. Pushing an aircraft beyond its flight envelope can damage the aircraft.

The flight envelope of an aircraft is determined by measuring the aircraft's handling performance at various speeds against the limiting load factor. For the purpose of calculating the flight envelope, speed is measured in terms of cruise speed, dive speed, and stall speed (the minimum speed required to sustain flight) under normal circumstances and during peak maneuvering. The **load factor** measures the stress on the aircraft in terms of its lift versus its weight.

Some more modern interpretations of the flight envelope include the following:

Extra Power

Extra power is a term used to describe the flight envelope, though it fails to consider the aircraft's actual performance at different speeds and during different maneuvers.

It is calculated by determining the power or thrust the aircraft requires to achieve straight and level flight and subtracting this number from the power the aircraft engine can generate. For example, if an aircraft requires 500 HP to sustain straight and level flight, and its engine can generate 1000 HP, the aircraft has 500 HP of extra power.

The extra power is available for the aircraft to perform flight maneuvers such as turning, climbing, and descending. Thus, the performance of the aircraft is determined by measuring the power necessary to perform these maneuvers based on the weight and required lift for the aircraft against the extra power available.

Doghouse Plot

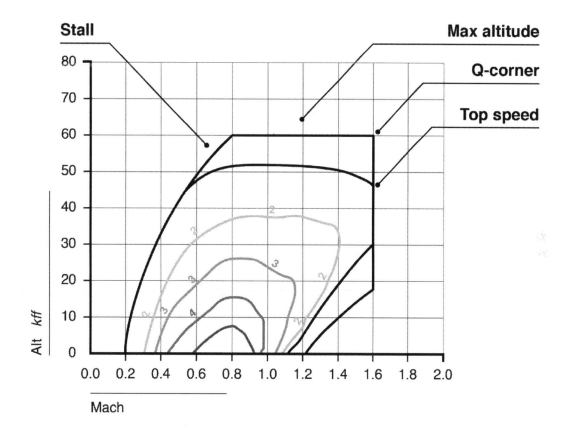

A **doghouse plot** involves a more complex set of calculations than the extra power method of determining the flight envelope. It provides a more complete and accurate representation of the aircraft's actual performance capabilities.

A doghouse plot describes the range of performance of which an aircraft is capable during turning and flying maneuvers, at different speeds and at different altitudes. A simplified doghouse plot visualizes the range between the stall speed (the minimum speed at which the aircraft can sustain flight) and the top speed at the maximum potential altitude in an upside-down U-shape. The data points inside the U-shape represent the aircraft's performance capability.

In this example, the doghouse plot shows the upward curve of the aircraft's performance as it begins to climb from the stall speed on x-axis. As the aircraft climbs, the speed increases as it accelerates to reach cruising speed during straight and level flight at the maximum altitude of 40,000. The U-shape begins to curve down when the aircraft reaches its maximum altitude, but speed increases as the aircraft loses altitude until it reaches top speed on the x-axis.

Airport Information

An airport is a central hub for the following commercial airline operations:

rr. Storing, maintaining, and servicing aircrafts
ss. Ensuring the aircrafts in their airspace operate safely
tt. Transporting passengers and crew

The landside of an airport includes the areas of the airport that are accessible to the public without passing through airport security. These areas include the roads leading to and from the airport, the parking lots, public transportation, and check-in areas. The airside of the airport includes the areas accessible after passing through airport security such as the taxiways, landing area, aprons, and boarding areas.

Features of an Airport

Landing Area
The landing area is the large, open space from which aircrafts take off and land. This area can refer to a runway, the deck of a ship large enough to allow an aircraft to take off and land, or even open water.

Runway
A runway or landing strip is a flat, paved or unpaved strip of land that provides enough length and width to allow airplanes to speed up before taking off and slow down upon landing. Runways feature runway thresholds that designate the different areas of the runway used for take-off and landing, and a runway safety area that functions like a shoulder on a freeway to reduce damage to an aircraft in the event of an accident. The blast pads of a runway are the flat surfaces marked with yellow chevrons that prevent erosion from jet blast.

Runway Lights
Runway lights are used to ensure safe landing at night. These lights include the following:

Runway end identifier lights: Flashing lights at the runway threshold

Runway end lights: Four lights on both ends of the runway that look green from the sky and red from the runway

Runway edge lights: White lights running the length of the runway on both sides

Runway centerline lighting system: Lights that run length-wise through the center of the runway at 50 ft. intervals. The lights are white until the last 3000 ft. of the runway, after which they alternate white and red until the last 984 ft., after which they are red.

Touchdown zone lights: White bars that run along either side of the centerline with three in each row for 3,000 ft.

Taxiway centerline lead-off and lead-on lights: Green and yellow alternating lights running along lead-off and lead-on markings along the centerline of the taxiway. The lead-off and lead-on lights direct traffic in opposite directions.

Land and hold short lights: White pulsating lights that designate the hold short position

Approach lighting system: Strobe lights and/or light bars that indicate the approach end, or the end of the runway

The lights are controlled by the control tower or flight service station, or they can be turned on by a pilot using pilot-controlled lighting (PCL) via radio.

Apron
The apron of an airport is the area outside the terminal or concourse where aircrafts are temporarily parked for refueling, loading and unloading cargo, and boarding and disembarking passengers. This area is often referred to as the "tarmac."

Control Tower
The control tower is a tall tower with a windowed command center from which Air Traffic Control (ATC) can monitor the air traffic in the local airspace.

Hangar
The hanger is the enclosed utility building in which aircrafts are stored when not in use to protect the aircrafts from environmental damage.

Terminal
The terminal is the area of the building that connects the landside of the airport to the airside. This area includes check-in and ticketing counters, ground transportation connections, and facilities for passengers. Travelers must pass through airport security to access the concourse, the area of the airport that contains the gates used to board and disembark from planes, as well as additional passenger facilities. The terms "terminal" and "concourse" are often used interchangeably.

Air Traffic Control (ATC)

Air Traffic Control is responsible for ensuring planes within the same airspace operate safely. ATC refers to both the people responsible for controlling air traffic and the ATC systems that manage safe operations at the airport. ATC includes the following duties:

The Flight Data Person receives pre-flight data from pilots, monitors the weather, transmits the flight plan to the FAA, and passes along the flight progress strip, the data used to track the progress of a flight.

Ground Control receives the flight progress strip to control aircraft traffic on the ground such as taxiing to or from the gates to the runway.

Local Control observes the sky and monitors radar from the control tower to track the movements of aircrafts in the local airspace.

Departure Control monitors radar and radio signals for the departing aircrafts at multiple airports from a Terminal Radar Control Approach facility.

Radar Control maintains communication between the aircraft and the ground within their designated sector and transmits instructions and information to the pilot.

Approach Control directs the pilot to the runway for landing and passes the flight back to local and ground control.

History of Aviation

Until the late 19th and early 20th century, humans failed to overcome the obstacles of heavier-than-air flight. Early attempts at flight included fashioning wings to allow humans to fly like birds. In 1485, Leonardo da Vinci designed the Ornithopter, upon which modern helicopters are based. The aircraft was never built.

Scientists turned to lighter-than-air flight. In 1709, Father Bartolomeu de Gusmão of Portugal presented the first model of a hot-air-balloon to the King John V. In 1783, the first hot air balloon, Montgolfière, was launched by Joseph-Michel and Jacques-Étienne Montgolfier in Lyon, France. The same year, Professor Jacques Alexandre César Charles and Les Frères Robert (the Robert Brothers) launched an unmanned hydrogen-balloon, Le Globe, in Paris.

The Montgolfier brothers continued to make progress that year, experimenting with animal passengers and, finally, in November, the brothers flew two human passengers in a hot air balloon. In December, Charles launched the first manned hydrogen balloon.

While lighter-than-air flight continued to progress, an English engineer named Sir George Cayley presented the concept of a fixed-wing aircraft in 1799. He built a winged glider that used the body motions of the pilot to steer. The aircraft had to be launched from a high surface to achieve lift because it had no means of propulsion to generate the necessary thrust, and it could not travel long distances. Cayley continued his studies into aerodynamics. He designed an aircraft that relied on the four forces of aerodynamics. Like modern aircrafts, it used airfoils to generate lift, controls to steer the craft, and a means of propulsion.

Despite these breakthroughs in the field of aviation, the problem of constructing an aircraft with an engine to generate enough thrust to lift an aircraft off the ground remained.

In the 19th century, German engineer Otto Lilienthal studied angles of attack to improve wing design. Early aviators made progress constructing helicopters and fixed-wing aircrafts with modern flight control surfaces that used steam-powered engines to generate thrust. These early aircrafts could successfully sustain heavier-than-air flight, but they could not travel long distances or carry people or cargo.

In 1903, after experimenting with fixed-wing aircrafts with his brother, Wilbur, in Kitty Hawk, NC, Orville Wright became the first person to successfully pilot an airplane, the Wright Flyer II, that used an engine and propellers to generate thrust, and two pairs of wings to generate lift. The Wright Brothers improved upon their design to achieve more controlled flight, and in October 1905, Wilbur sustained a flight of 24.2 miles in the redesigned Flyer III. The Wright Brothers' work represented a breakthrough in the field of aviation, and the field continued to progress rapidly during World War I as world leaders rushed to militarize aviation. In 1919, the U.S. Navy ship, Curtiss NC-4, made the first transcontinental flight from Newfoundland to Portugal, and in 1927, Charles Lindbergh flew across the Atlantic Ocean in a 33.5-hour flight. Since then, the field of aviation has continued to progress, and after the invention of the jet in the 1950s, flight became a common means of travel.

Practice Questions

1. Which of the following aerodynamic forces limits flight?
    a. Lift
    b. Drag
    c. Weight
    d. Thrust
    e. Power

2. Which type of motion occurs when the aircraft is banked?
    a. Yaw
    b. Roll
    c. Pitch
    d. Spin
    e. Stall

3. Which of the following terms is defined as a surface that generates lift by splitting oncoming airflow into faster moving air above the surface and slower moving air below the surface?
    a. Chord line
    b. Angle of attack
    c. Airfoil
    d. Lift coefficient
    e. Chord length

4. Which of the following types of stress causes an object to stretch?
    a. Torque
    b. Tension
    c. Compression
    d. Shear
    e. Bending

5. Which of the following components provides length-wise support to the fuselage frame?
    a. Former
    b. Bulkhead
    c. Truss
    d. Longeron
    e. Monocoque

6. Which of the following components provides the main structure of a wing?
    a. Airfoil
    b. Wing spar
    c. Wing root
    d. Bulkhead
    e. Wing skin

7. Which of the following flight control surfaces is attached to the horizontal stabilizer?
    a. Rudder
    b. Aileron
    c. Elevator
    d. Wing
    e. Boom

8. Which of the following parts of an aircraft often houses the engine and other equipment?
    a. Cowling
    b. Nacelle
    c. Fairing
    d. Booms
    e. Fuselage

9. Which of the following movements will occur if front pressure is applied to the elevator?
    a. The nose of the aircraft will go up
    b. The left wing will go up
    c. The left wing will go down.
    d. The aircraft will turn to the right on its vertical axis
    e. The nose of the aircraft will go down

10. Which of the following should the pilot adjust to achieve a straight and level flight?
    a. Pitch and roll
    b. Yaw and pitch
    c. Roll and yaw
    d. Just pitch
    e. Just yaw

11. Which of the following types of motion does the pilot adjust to steer the helicopter?
    a. Spin
    b. Yaw
    c. Roll
    d. Pitch
    e. Rotor

12. Which of the following components of a helicopter transmits the input from the pilot to the rotors in order to steer the aircraft?
    a. Transmission
    b. Mast
    c. Swash plate
    d. Collective pitch lever
    e. Cyclic pitch lever

13. What is the purpose of the tail rotor?
    a. To generate lift
    b. To counter torque from the main rotor
    c. To adjust the pitch
    d. To stabilize the helicopter
    e. To help the helicopter yaw

14. The flight envelope is measured in terms of which of the following?
    a. Horsepower vs. weight
    b. Speed vs. load factor
    c. Load factor vs. maximum altitude
    d. Stall speed vs. top speed
    e. Top speed vs. horsepower

15. Which of the following methods calculates the flight envelope as the difference between the power required to maintain straight and level flight, and the maximum potential power of the aircraft?
    a. Performance envelope
    b. Extra power
    c. Load factor
    d. Doghouse plot
    e. U-shaped plot

16. Which of the following is the area of an airport that connects the landside to the airside?
    a. Terminal
    b. Runway
    c. Taxiway
    d. Concourse
    e. Apron

17. Which of the following is area often referred to as the "tarmac"?
    a. Terminal
    b. Hangar
    c. Apron
    d. Runway
    e. Concourse

18. Which of the following air traffic control roles is responsible for directing the pilot to the runway for landing?
    a. Local Control
    b. Ground Control
    c. Departure Control
    d. Approach Control
    e. Radar Control

19. Which of the following early aviators experimented with angles of attack to improve wing design?
    a. Orville Wright
    b. George Cayley
    c. Charles Lindbergh
    d. Leonardo da Vinci
    e. Otto Lilienthal

20. Which of the following early aviators designed a fixed-wing aircraft with an engine and propeller and two pairs of wings?
    a. The Wright Brothers
    b. The Robert Brothers
    c. George Cayley
    d. Otto Lilienthal
    e. The Montgolfier Brothers

Answer Explanations

1. C: Weight limits flight, as gravity constantly pulls an aircraft back down to the ground. As such, the thrust (Choice *D*), which is sometimes called power (Choice *E*), must be sufficient to propel the weight of the aircraft off the ground to achieve lift (Choice *A*), which is the force that keeps the aircraft in the air. Drag (Choice *B*) is a resisting, but not a limiting, force. It runs parallel to the object in motion, trying to push the object back, while weight pulls the object down.

2. B: When the right or left airplane wing is banked or lowered, the aircraft will roll or turn in the direction of the lowered wing. When an airplane yaws (Choice *A*), it spins left or right on its vertical axis, and when it pitches (Choice *C*), it tilts up or down on its lateral axis. Spin (Choice *D*) is a type of motion that occurs as the result of an improperly corrected stall. Stall (Choice *E*) occurs when a sudden decrease in lift and increase in drag causes the aircraft to fall.

3. C: The wing of an aircraft is an airfoil surface that is shaped to generate lift. The lift coefficient (Choice *D*) is determined by the angle of attack (Choice *B*), which is measured as the angle between the chord line (Choice *A*) and the direction of the oncoming airflow. The chord length (Choice *E*) is the distance of the chord line from the leading edge to the trailing edge.

4. B: Tension is a pulling force that causes an object to stretch, while compression (Choice *C*) is the opposite force that causes objects to reduce in size. Shear (Choice *D*) is the force that runs parallel to an object that shears or breaks off layers of the object. Finally, torque (Choice *A*) twists an object and bending (Choice *E*) bends an object.

5. D: A longeron provides longitudinal support to a truss (Choice *C*) frame assembly. A former (Choice *A*) provides the shape, and bulkheads (Choice *B*) resist pressure in a monocoque frame assembly. A semimonocoque frame assembly includes features of both truss and monocoque frame assemblies, including a former, bulkheads, and longeron. A monocoque frame assembly is a single-shell frame that relies on the strength of its skin or covering to withstand stress. A monocoque frame assembly uses formers that determine the shape of the fuselage and bulkheads to resist pressure.

6. B: Wing spars are supports that run from wing root (Choice *C*) to wing tip to provide the primary support and structure for a wing. Bulkheads (Choice *D*) are used to resist pressure in a fuselage, while the airfoil (Choice *A*) is the wing itself. Wing skins (Choice *E*) help support the wing spars.

7. C: The elevator is attached to the horizontal stabilizer, or tail wing, to control the pitch, while the rudder (Choice *A*) is attached to the vertical stabilizer to control the yaw. The ailerons (Choice *B*) are attached to the wing (Choice *D*), or airfoil, to control the roll that turns the aircraft. Booms (Choice *E*) are not flight control surfaces. They contain fuel tanks, extend the tail, or provide additional support.

8. B: A nacelle is a compartment that may be attached to the fuselage or to the wing in which the engine, engine equipment, and landing gear is often stored. The cowling (Choice *A*) is the removable panel on a nacelle that slides back to provide access to the engine or when the landing gear is engaged. A fairing (Choice *C*) smooths the edges on the cowling or other areas of the aircraft to reduce drag. Booms (Choice *D*) are extensions of the fuselage upon which a nacelle could be mounted. The fuselage (Choice *E*) is the main supporting structure of the aircraft that carries the pilot, passengers, and cargo.

9. E: When the pitch is adjusted by applying front pressure to lower the elevator, the nose goes down and the tail goes up. Applying back pressure to raise the elevator will cause the tail to go down and the nose to go up (Choice *A*). The motion of the wings (Choices *B* and *C*) are controlled by the ailerons, and the rudder spins the aircraft on its vertical axis (Choice *D*).

10. A: To achieve straight and level flight, the pilot must adjust for both the changes in roll, so the aircraft heading remains straight, and pitch, so the aircraft remains level in relation to the horizon. Roll and yaw (Choice *C*) are used in combination to turn the aircraft, while a sudden shift in pitch will cause the aircraft to yaw. Adjusting the pitch alone (Choice *D*) will fail to maintain a straight heading. Just yaw (Choice *E*) will not achieve straight and level flight as yaw is the rotation of the aircraft around its vertical axis.

11. D: The pilot uses the cyclic and collective pitch levers to steer a helicopter. Spin (Choice *A*), yaw (Choice *B*), and roll (Choice *C*) are not employed in the steering of a helicopter. The rotor (Choice *E*) is not a type of motion; it is comprised of an array of small airfoils (rotor blades) that spin constantly around a shaft.

12. C: The lower swash plate receives input from the pilot through control rods connected to the pitch levers the pilot uses to steer while the upper swash plate is connected to the mast (Choice *B*) and spins to transmit the input from the pilot through control rods on the rotor blades. The transmission (Choice *A*) is used to transmit power to the rotor from the engine. Choice *E*, the cyclic pitch lever, controls the helicopters sideways and the forward-and-backward motion by tilting the angle of the pitch. The collective pitch lever (Choice *D)* controls the helicopter's vertical motion by increasing or decreasing the pitch.

13. B: The purpose of the tail rotor is to enable steering by countering the torque from the main rotor. An airfoil generates lift (Choice *A*), while the pitch levers control pitch (Choice *C*). The helicopter remains stable (Choice *D*) due to a stabilizer bar mounted on top of the main rotor blades. Choice *E* is incorrect because the tail rotor helps prevent yaw.

14. B: The flight envelope is the minimum and maximum range of the performance of the aircraft as measured by its potential speed vs. its limiting load factor. Weight (Choice *A*) contributes to the measure of the load factor while the extra power method subtracts the horsepower (Choice *A*) required to fly straight and level from the maximum potential power of the aircraft. A basic doghouse plot defines the range of performance between the stall speed and top speed (Choice *D*) at maximum altitude (Choice *C*). Choice *E* is not reflected by the flight envelope.

15. B: The extra power method of calculating the flight envelope subtracts the used power from the potential power, against which it measures the airplane's ability to perform turning, climbing, and descending maneuvers. Performance envelope (Choice *A*) is an alternative term for flight envelope, and the load factor (Choice *C*) is a metric used to calculate the flight envelope. A doghouse plot (Choice *D*) shows a range of performance in an upside-down U-shaped plot (Choice *E*).

16. A: The terminal is the connection point between the landside and airside of an airport. Travelers and staff must pass through airport security to reach the concourse (Choice *D*) on the airside that contains the boarding gates. The runway (Choice *B*) is in the landing area of the airside on which planes take-off and land. The taxiway (Choice *C*) connects the apron to the runway. Choice *E*, the apron, is where planes park temporarily for refueling, boarding and disembarking passengers, and loading and unloading.

17. C: Although the apron (where planes park temporarily for refueling, boarding and disembarking passengers, and loading and unloading) is typically constructed of concrete, it is often referred to as the "tarmac." The terminal (Choice *A*) is the building that connects the landside to the airside of the airport, while the hanger (Choice *B*) is the building in which aircrafts are stored to protect them from the elements. The runway (Choice *D*) is a paved or unpaved strip of land used to take off or land aircrafts. The concourse (Choice *E*) is the area of the airport that contains the gates used to board and disembark from plane.

18. D: Approach Control is responsible for directing pilots to the runway when they approach for landing. When the plane is on the ground, Ground Control (Choice *B*) takes over to direct the pilot from the runway to the apron. Local Control (Choice *A*) monitors the local airspace for incoming and departing flights, while Departure Control (Choice *C*) monitors and directs flights departing the airport. Choice *E*, Radar Control, maintains communication between the aircraft and the ground within their designated sector and transmits instructions and information to the pilot.

19. E: In 1870, Otto Lilienthal improved wing design on his glider by experimenting with different angles of attack to generate lift. His research was later used by Orville and Wilbur Wright (Choice *A*) in the design of their airplane in 1903. Sir George Cayley (Choice *B*) is responsible for the discovery of the forces of aerodynamics and the design of a fixed-wing aircraft with airfoils, steering controls, and a means of propulsion in the early 1800s. Charles Lindbergh (Choice *C*) flew across the Atlantic Ocean in a 33.5-hour flight. In the late 1400s, Leonardo da Vinci (Choice *D*) designed a rotary-motion aircraft upon which helicopters are based.

20. A: The Wright Flyer II, a biplane with an engine and front propeller, was the first fixed-wing aircraft piloted by a human that successfully achieved lift off, controlled flight, and landing in 1903. It was invented by the Wright Brothers. The Robert Brothers (Les Frères Robert) (Choice *B*) launched Le Globe, an unmanned hydrogen balloon in 1783. Sir George Cayley (Choice *C*) is credited with discovering aerodynamics in the early 1800s, Otto Lilienthal (Choice *D*) improved wing design in the late 1800s, and the Montgolfier Brothers (Choice *E*) were the first to have a manned hydrogen balloon.

Greetings!

First, we would like to give a huge "thank you" for choosing us and this study guide for your AFOQT exam. We hope that it will lead you to success on this exam and for your years to come.

Our team has tried to make your preparations as thorough as possible by covering all of the topics you should be expected to know. In addition, our writers attempted to create practice questions identical to what you will see on the day of your actual test. We have also included many test-taking strategies to help you learn the material, maintain the knowledge, and take the test with confidence.

We strive for excellence in our products, and if you have any comments or concerns over the quality of something in this study guide, please send us an email so that we may improve.

As you continue forward in life, we would like to remain alongside you with other books and study guides in our library. We are continually producing and updating study guides in several different subjects. If you are looking for something in particular, all of our products are available on Amazon. You may also send us an email!

Sincerely,
APEX Test Prep
info@apexprep.com

FREE

Free Study Tips DVD

In addition to the tips and content in this guide, we have created a FREE DVD with helpful study tips to further assist your exam preparation. **This FREE Study Tips DVD provides you with top-notch tips to conquer your exam and reach your goals.**

Our simple request in exchange for the strategy-packed DVD is that you email us your feedback about our study guide. We would love to hear what you thought about the guide, and we welcome any and all feedback—positive, negative, or neutral. It is our #1 goal to provide you with top-quality products and customer service.

To receive your **FREE Study Tips DVD**, email freedvd@apexprep.com. Please put "FREE DVD" in the subject line and put the following in the email:

a. The name of the study guide you purchased.

b. Your rating of the study guide on a scale of 1-5, with 5 being the highest score.

c. Any thoughts or feedback about your study guide.

d. Your first and last name and your mailing address, so we know where to send your free DVD!

Thank you!

Made in the USA
Middletown, DE
22 February 2020